FUNDAMENTALS OF PLANT PATHOLOGY

NIPA GENX ELECTRONIC RESOURCES & SOLUTIONS P. LTD.
New Delhi-110 034

About the Author

Dr. Sanjeev Kumar is Assistant Professor/Scientist, Department of Plant Pathology, Jawaharlal Nehru Krishi Vishwavidyalaya, Jabalpur, Madhya Pradesh. He is very sincere and hard working teacher. He has taught many undergraduate and postgraduate courses with great dedication to the great satisfaction of the students and is actively involved in all teaching activities. Dr. Kumar has guided ten M.Sc student and has been actively involved in his research and extension activities too. As a researcher, he has been associated with 9 major research projects sponsored by ICAR, NATP, TSP and JICA, Japan. He has published 66 research and reviews papers, 23 book chapters, 110 popular articles, 1 research bulletin and 3 technical folders. He is an active member of over 6 National and International Societies. He is Fellow of Indian Phytopathological Society and Society of Biocontrol Advancement. He has attended many conferences, symposia, workshops and presented papers. He has also attended the World Soybean Research Conference held in Durban, South Africa.

Dr. Kumar has authored five books entitled *Plant Pathogens & Principles of Plant Pathology*, *Diseases of Horticultural Crops: Identification & Management*, *Diseases of Field Crops and Their Integrated Management*, *Pesticides and Plant Protection Appliances*, Published by New India Publishing Agency, Pitam Pura, New Delhi, Textbook *of Diseases of Field & Horticultural Crops & Their Management-I*, Published by Brillion Publishing, Karol Bagh New Delhi-10005 and three practical manuals. His research interests include pulse pathology and biological control.

FUNDAMENTALS OF PLANT PATHOLOGY

Sanjeev Kumar
Assistant Professor/Scientist
Department of Plant Pathology
Office of Dean, Faculty of Agriculture
Jawahar Lal Nehru Krishi Vishwavidyalaya
Jabalpur, Madhya Pradesh-482004, India

NIPA GENX ELECTRONIC RESOURCES & SOLUTIONS P. LTD.
New Delhi-110 034

NIPA GENX ELECTRONIC
RESOURCES & SOLUTIONS P. LTD.
101,103, Vikas Surya Plaza, CU Block
L.S.C.Market, Pitam Pura, New Delhi-110 034
Ph : +91 11 27341616, 27341717, 27341718
E-mail:newindiapublishingagency@gmail.com
www: www.nipabooks.com

For customer assistance, please contact
Phone: + 91-11-27 34 17 17 Fax: + 91-11- 27 34 16 16
E-Mail: feedbacks@nipabooks.com

ISBN : 978-93-90591-20-6

Composed and Designed by NIPA.

Preface

The Indian Council of Agricultural Research (ICAR), New Delhi an apex body developed new *course* curricula for UG and PG students under the umbrella of **Fifth Dean's Committee** keeping in view the requirement and maintain the pace with industrial needs to increase the farmers income as per programme of **Prime Minister of India**. The undergraduate syllabus has been completely revised and has rendered existing textbooks obsolete or incomplete for students. This gap in knowledge give me an excellent opportunity to present updated textbooks to students as per their syllabus requirements. The purpose of writing this text books is to provide the basic and emerging facts whereby the students may be introduced to the scientific foundation of this basic course Fundamentals of Plant Pathology.

The book aims to develop in the young minds the passion for learning and analysing the environment, to probe into the problems and think of the possible solution. While preparing the book, I have taken utmost care to present facts according to the level of our students and graded our content accordingly.

I wish to express my sincere gratitude to our past and present pioneers in the field of Plant Pathology whose work has been freely used in the prepare of this book. I am also indebted to my Parent in laws, wife Dr. (Mrs) Archana Rani, children Saumya and Adyan for bearing with me during the preparation of this book .

Omission and mistakes are bound to be found in this endeavor and I am prepared to bear the accountability of making any misrepresentation of facts. Teachers and students are requested to comment and give suggestions for further improvements in the book We expect the book to set the students on a pursuit where they learn more by asking than by being told.

June, 2020

Sanjeev Kumar
Jabalpur

Contents

Glossary

Acervulus – A saucer-shaped, spore-producing body of a fungus embedded in host tissue.

Aquisition period – The period of time need for a virus vector to aquire a virus while feeding on an infected plant host.

Actinomycetes – Filamentous bacteria that produce several antibiotics and give soil its earthy smell.

Active ingredient – In pesticides, the chemical responsible for the desired effect.

Aggressiveness – Virulent forms of pathogen cause differing degrees of symptom severity.

Alternate Host – One of two kinds of plants on which a parasitic fungus (e.g., rust) must develop to complete its life cycle.

Anamorph – Asexual stage of a fungus.

Antagonism – The counteraction between organisms or groups of organisms

Antibiotic – A complex chemical substance produced by one microorganism that inhibits or kills other microorganisms (e.g., streptomycin).

Antibody – A specific protein formed in the blood of warm-blooded animals in response to the injection of an antigen.

Antigen – Any foreign chemical (normally a protein) that induces antibody formation in animals.

Antiseptic – A substance that prevents, retards, or destroys microorganisms.

Apothecium – An open, cuplike, or saucer-shaped sexual fungal fruiting body containing asci.

Autoecious – The need of only one host for completing the life cycle of a rust.

Avirulence – The inability of a pathogen to cause disease.

Avr gene – Avirulence gene, a pathogen gene whose product is recognized by a plant and leads to a resistant reaction in the plant.

Bactericide – A compound toxic to bacteria.

Biotroph – A plant pathogenic fungus that requires living host cells i.e. an obligate parasite.

Biotype – The smallest morphological unit within a species, the members of which are usually genetically identical.

Carrier – A plant or animal that carries a virus or other infective agent without showing symptoms.

Chemotherapy – Treatment of disease by chemicals working internally. Chemical agent has toxic effect directly or indirectly on the pathogens without injury to the host plant.

Circulative transmission – Vector trasmission of a virus where virus particles must circulate through the vectors hemolymph and enter the salivary glands to be transmitted.

Cleistothecium – Closed, usually spherical, ascus-containing structure of powdery mildew fungi. A sexual fruiting structure.

Collateral host – The wild host of same families of a pathogen.

Curative – Chemical control method aimed at inhibiting the development of an established infection.

Diagnostic – A distinguishing characteristic serving to identify or determine the presence of a disease or other condition.

Differential hosts – Host range can be tested and compared to known pathogens for identification.

Disease – Any deviation in the general health, or physiology or function of plant or plant parts, is recognized as a disease.

Disease cycle – The chain of events involved in disease development.

Disease incidence – The number or poroportion of individuals that are diseased.

Disease-gradient curve – A graph that plots disease vs. distance from an inoculum source.

Disease progress curve – A graph that plots disease vs. time.

Disease severity – The amount or proportion of tissue that is diseased.

Disease syndrome – The set of varying symptoms characterizing a disease.

Disinfectant – Any agent for destroying the causal agent of disease after infection.

Disinfestant – Any agent that removes, kills, or inactivates disease-causing organisms *before* they can cause infection.

Dissemination – The spread of infectious material (inoculum) from a diseased to a healthy plant by wind, water, humans, insects, animal, machinery, or other means.

DNA fingerprinting – Technique for pathogen identification where restirction enzymes are used to cut genomic DNA, producing a pattern specific to a pathogen.

Dormancy – Nongrowing (inactive, quiescent) state of a plant.

ED 50 – Effective dose, the amount needed to have the desired effect in 50% of the population.

Ectoparasite – A parasite that feeds from outside the host.

ELISA – A test used to detect antigens. Antibodies can be created to detect pathogen proteins.

Endoparasite – A parasite that enter and feed from inside the host.

Endogenous – Produced inside.

Endophytic – Living within another plant.

Enphytotic – Plant disease that causes about the same amount of injury each year.

Environment – The external conditions and influences that surround living organisms.

Epidemiology – The study of factors influencing the initiation, development, and spread of infectious disease.

Epiphytotic – The widespread and destructive development of a disease on many plants in a community or communities.

Eradicant – Chemical used to eliminate a pathogen from a host or an environment.

Eradication – Control of disease by eliminating the pathogen after it is already established.

Escape – Plants in a given population that remain free of disease where it is prevalent, although they possess no natural inherent resistance to the disease.

Etiolation – Yellowing and long, spindly growth as a result of insufficient light.

Etiology – The description of the cause of disease.

Exclusion – Control of disease by preventing its introduction into disease-free areas.

Exogenous – Produced outside.

Exudate – A substance (usually liquid) formed inside a plant and discharged from diseased or injured tissue.

Facultative Parasite – An organism that is ordinarily saprophytic but under proper conditions may be parasitic.

Facultative Saprophyte – An organism that is ordinarily parasitic but under proper conditions may be saprophytic.

Fission – Bacterial reproduction by simple cell division.

Flagellum – A long hairlike or whiplike contractile filament protruding from certain bacterial cells and spores of fungi and that enable movement.

Fruiting Body – Any of various complex, spore-bearing fungal structures.

Fumigant – Vapor-active chemical used in the gaseous phase to kill or inhibit the growth of microorganisms.

Fungi Imperfecti – A major group of fungi for which no sexual production of spores is known.

Fungicide – An agent that inhibits or kills fungi.

Fungistat – A chemical or physical agent that prevents fungi from developing but does not kill them.

Fungus – A single or many-celled, naked or covered, irregular or filamentous organism, usually with a chitinous cell wall. Lacking chlorophyll and incapable of manufacturing its own food, it feeds on dead or living plant or animal matter.

Gall – An unusual enlargement on some portion of a plant.

Giant Cells – Large, usually multinucleate cells formed by abnormal cell fusions or failure of proper cell wall formation following growth and nuclear division.

Haploid – The chromosome number of the gametophytic generation or phase or having a single complete set of chromosomes.

Haustorium – A modified mycelial branch that grows into a plant cell, makes intimate contact with the protoplast, and absorbs food.

Hemibiotroph – A plant pathogenic fungus that initially requires living host cells but after killing the host cell grows on the dead and dying cells.

Heteroecious – Requiring two or more unrelated hosts for completing the life cycle of a rust.

Heterothallic – Producing fusing gametes on separate and distinct mycelia.

Hemiparasites – Parasitic plants that have chlorophyl and can make some of their own sugar but still rely on thier host for water and other nutrients.

Holoparasites – Parasitic plants that do not have chlorophyl and rely on thier host for all nutrients.

Homothalic – Producing fusing gametes on the same mycelium.

Host – The plant on or in which a parasite lives and from which it obtains its food.

Hyaline – Clear, translucent.

Hypersensitive response – Plant responds to pathogen infection by quickly killing the infected cells, blocking the advance of the pathogen.

Hyperplasia – A symptom due to an abnormal increase in the number of individual cells.

Hypertrophy – An symptom due to an abnormal increase in the size of individual cells.

Hyphae – Fungal filaments which collectively form the mycelium of a fungus.

Hypoplasia – The underdevelopment of cells, tissues or organs.

Immune – Cannot be infected by a given pathogen.

Immunity – A relationship between a plant and a causal agent in which the plant does not become diseased.

Incubation period – The period of time between penetration of a pathogen to the host and the first appearance of symptoms on the plant.

Indexing – Determining presence of disease in a plant by removing buds or other parts for inoculation of a susceptible indicator plant that exhibits specific symptoms of a transmissible disease.

Infection – The initiation and establishment of a parasite within a host plant.

Infection court – Specific area on a plant where a pathogen gains entrance to the host.

Infest – To overrun or contaminate.

Infestation – Presence in numbers (e.g., of insects, mites, or nematodes).

Indicator plants – Certain plant species or cultivars that show characteristic symptoms of infection.

Inoculum – That portion of pathogen which is transferred to plant & cause disease.

Inoculation – The process of transferring inoculum to host.

Inoculum density – The number of infective units in a given volume or area.

Inoculum potential – The growth or threshold of fungus available for colonization at host.

Invasion – The penetration and spread of a pathogen in the host.

Intercellular – Between the cells.

Intracellular – Within the cells.

Juvenile – Immature, non-reproductive stage of nematodes.

Latent – Present but not manifest or visible, as a symptomless infection.

Lesion – A local injury or delimited diseased area.

Local invasion – That involving only a portion of the plant.

Local necrosis – The death or disintegration of cells and tissues in a localized area of an organ.

Macrocyclic – Rusts that produce all five spore types.

Microcyclic – Rusts that lack one or more of the five spore types.

Macroscopic – Visible to the naked eye, without the aid of a microscope.

Microscopic – Visible only with the aid of magnification.

Migratory – Describes a life style of plant parasitic nematodes that move through the host as they feed.

Mechanical transmission – Virus transmission from plant to plant by infected plant sap.

Monocyclic disease – A disease where only one disease cycle is completed each year or growing season.

Monotrichous – Having only one flagellum.

Mosaic – Disease symptom characterized by nonuniform foliage coloration, with a more or less distinct intermingling of normal and light green or yellowish patches.

Multiple cycle disease (Polycyclic) – Some pathogens specially a fungus, can complete a number of life cycles within one crop season of the host plant and the disease caused by such pathogens is called multiple cycle disease e.g. wheat rust, rice blast, late blight of potato etc.

Mushroom – A conspicuous fleshy fungus fruiting body.

Mutualism – Symbiosis of two organisms that are mutually helpful or that mutually support one another.

Mycelium – The mass of interwoven threads (hyphae) making up the vegetative body of a fungus.

Mycoplasma – Degenerate bacteria that do not have cell walls.

Mycorrhiza – A symbiotic association of a fungus with the roots of a plant.

Necrosis – A symptom marked by rapid death of the host or parts of the host.

Necrotroph – A pathogenic fungus that kills the host and survives on the dying and dead cells.

Nematicide – A chemical or physical agent that kills, inhibits, or protects against nematodes.

Nematodes – Generally microscopic tubular worms, usually living free in moist soil, water, and decaying matter, or as parasites of plants and animals.

Non-persistant transmission – Vector trasmission of a virus where the vector quickly picks up virus particles on its mouthparts and is infective for a short period of time.

Obligate parasite – An obligate parasite is an organism that can live only on living tissue.

Oogonium – Female egg cell of oomycete fungi.

Oomycete – A group of fungi that produce oospores such as *Pythium, Phytophthora* and *Aphanomyces.*

Oospore – Thick-walled, sexually-derived resting spore of oomycete fungi.

Overwinter – To survive over the winter period.

Parasite – An organism that lives within or upon another living organism from which it derives nourishment and in which it may cause various degrees of injury.

Parasitism – The phenomenon of the growth of one organism, the parasite, at the expense of another, the host.

Pathogen – An entity, usually a micro-organism that can cause the disease.

Pathogenicity – An entity's capacity for producing a disease.

Pathogenesis – It is a process caused by an infectious agent (pathogen) when it comes in contact with a susceptible host.

Pathology – The study of disease.

PCR – technique to amplify sequences of DNA. Can be used to detect specific sequences of pathogen DNA.

Pectinase – The enzyme that breaks down pectic substances to simple carbohydrates.

Perithecium – A round to flask-shaped, thick-walled spore case containing asci and with an ostiole.

Peritrichous – Having flagella all over the outside of the cell.

Pesticide – Any chemical or physical agent that destroys pests.

Phage – Virus that attacks bacteria.

Phyllody – Change from a normal flower to leafy structures.

Physiogenic disease – A disease produced by some unfavorable physical or environmental factors.

Physiologic race – One or a group of microorganisms similar in morphology but dissimilar in certain cultural, physiological or pathological characters.

Phytoplasma – Microorganisms found in phloem tissue that resemble mycoplasmas in all respects except that they cannot yet be grown on artificial nutrient media.

Phytotoxic – Injurious to plants.

Plasmodium – A naked, multinucleate, vegetative body capable of amoeboid motion.

Polymorphism – The existence of several asexual spore stages in the life cycle of an organism.

Predisposition – The effect of one or more environmental factors which makes a plant vulnerable to attack by a pathogen.

Preventive – Chemical control method aimed at preventing infection of the pathogen.

Primary infection – The first infection of a plant by the over wintering or over summering of the pathogen.

Primary Inoculum – Inoculum, usually from an overwintering source, that initiates disease in the field, as opposed to inoculum that spreads disease during the season.

Propagule – The part of an organism that may be spread so as to reproduce the organism.

Propagative transmission – Vector trasmission of a virus where virus particles are repliacted in the the vector. In this case the vector retention time is longer than with circulative transmission and, in some cases, the virus can be transovarially transmitted.

Prophylaxis – Methods used to preserve health and prevent spread of disease.

Protectant – A chemical applied to a plant surface in advance of the pathogen to prevent infection.

Pustule – A local elevation of the epidermis that may rupture to expose the causal agent.

Pycnidium – The asexual, globose or flask-shaped fruiting body of fungi-producing conidia.

Quarantine – Regulation forbidding sale or shipment of plants or plant parts, usually to prevent disease, insect, nematode, or weed invasion of an area.

Race – A strain of a pathogen characterized by the limitation of its host range to certain species and varieties of plants.

Resistance – The sum of the qualities of the host and causal agent that retard the activities of the causal agent.

Rhizoid – Intercellular thallus branch that absorbs food and provides anchorage.

Rhizomorph – An aggregation of hyphae into a cordlike or rootlike strand.

Rickettsia – A single-celled animal and human disease--causing organism with a partial cell wall that has not been grown in culture.

Rust – A particular fungus, any of its stages or the disease caused by any of the stages.

Sanitation – Destroying all infested and infected plant parts during the season.

Saprophyte – An organism that derives its nourishment from dead organic matter.

Scab – Crustlike disease lesion.

Secondary Infection – Infection resulting from the spread of infectious material produced after a primary infection.

Sedentary – Describes a life style of plant parasitic nematodes that stay in one place and set up a feeding site.

Semi-persistant transmission – Vector trasmission of a virus where virus particles enter the vectors foregut. In this case the vector also picks up the virus quicly but is infective for somewhat longer than with non-persistant

Sign – The structure of the pathogen itself.

Single cycle disease (Monocyclic) – This type of disease is referred to those caused by the pathogen that can complete only one life cycle in one crop season of the host plant. e.g. downy mildew of rapeseed, club root of crucifers, sclerotinia blight of brinjal etc.

Soilborne – Fungi able to survive in the soil as saprophytes. Also called "soil inhabitant."

Spiroplasma – A single-celled, wall-less, spiral, filamentous organism associated with corn stunt and citrus stubborn disease.

Sporangiophore – A sporangium-bearing hypha.

Sporangium – A fruiting body that produces asexual spores within a more or less spherical wall.

Spore – A fungal reproductive unit or seed that serves as an agent of dispersal and propagation.

Sporodochium – A cushion-shaped spore-producing body of a fungus.

Sporogenous – Capable of forming spores.

Sporulation – The process of producing spores.

Sterilant – Any agent or chemical that destroys all living organisms in a substance such as soil.

Stomata – A small pore or opening in the epidermis of leaves and stems through which gases pass.

Stunt – A wide range of parasitic and nonparasitic agents.

Stunted – An unthrifty plant reduced in size and vigor due to unfavorable environmental conditions.

Stylet – Slender, tubular mouthparts in plant-parasitic nematodes or aphids.

Substrate – The substance or object on which an organism lives and from which it gets nourishment.

Susceptibility – The sum of the qualities of a plant and causal agent that allows the development of the causal agent.

Symbiosis – A mutually beneficial association of two or more different kinds of organisms.

Symptoms – The external and internal reaction or alterations of a plant as a result of disease.

Systemic – Pertaining to a disease in which an infection leads to general spread throughout the plant body. Also, a chemical that spreads internally through a plant.

Systemic aquired resistance – Whole plant resistance response that occurs following an earlier localized exposure to a pathogen.

Teliomorph – Sexual stage of a fungus.

Teliospore – Thick-walled resting spore produced by some fungi, notably rusts and smuts, that germinates to form a basidium.

Thallus – The vegetative body of the lower plant that has not differentiated into stems and leaves.

Ti-plasmid – Tumor inducing plasmid of *Agrobacterium tumefaciens* that causes gall formation and opine production by infected plants.

Tolerance – Ability of the plant to endure the development of the parasite without showing marked symptoms of disease.

Transgenic plants – Plants that have been genetically manipulated to express a gene from a different species.

Transovarial transmission – Viral transmission from an insect vector to its offspring, meaning offspring of an infected vector are also infective.

Transposons – A segment of chromosmal DNA that can move around in the genome.

Tumor – A swelling or protuberance

Tylosis – A bladderlike intrusion of the protoplasm from a parenchymatous cell through a pit into the lumen of a xylem cell.

Vector – An agent, such as an insect, nematode, or fungus, that may transmit a pathogen.

Viroid – An infectious nucleic acid without a protein coat that causes potato spindle tuber or chrysanthemum stunt.

Virulence – The degree of infectivity of a given pathogen.

Viruliferous – Capable of transmitting a virus.

Virus – Submicroscopic, infectious agent, too small to be seen with a compound microscope, that multiplies only in living cells. A virus consists of nucleic acid surrounded by a protein coat.

Water-soaked – Describing plants or lesions that appear wet and dark and are usually sunken and translucent.

Wilt – Lack of freshness and turgor and drooping of leaves from lack of water; a vascular disease that interrupts the plant's normal uptake and distribution of water.

Witches' Broom – Abnormal, brush-like development of many weak shoots.

Yellowing – The yellow color of plant parts resulting from the excessive proportion of yellow pigments, in turn produced by the underdevelopment or partial destruction of the green pigments.

Yellows – A disease characterized by yellowing and stunting of affected parts (caused by fungi, virus, bacteria, or deficiency of essential elements).

Zoospore – Fungus spore with flagella, able to move in water.

Zygospore – A fungal resting spore produced by the fusion of equal gametes.

1

Introduction, Importance of Plant Diseases, Scope and Objectives of Plant Pathology

Of the 7 billion global population, about 800 million do not have enough food to eat and majority of the them live in developing countries. By 2050, the world population is estimated to reach 9 billion. Plant pests and diseases are responsible for losses of 20 to 40% of global food production. Due to increasing global trade and a changing climate, plant diseases pose a major threat to food security today than ever before. Agriculture will have to cope with weather fluctuations that can affect the survival, distribution and multiplication of plant pathogens apart from shift in pathogenic microbial flora. Emergence of new races of pathogen potentially threaten the production of food. To cite an example, migration of 13 -A2 clonal lineage of *Phytophthora infestans* from Europe, during 2013, devastating tomato and potato in India. Plant pathogens also impact on environment sustainability, forest health, recreational horticulture, human and animal health.

The science of plant pathology can play a much needed role in alleviating hunger and malnutrition through development of affordable solutions for devastating plant diseases leading to increase in food production. It hold promise to ensure quality produce by containing microbial invasions and toxin contamination of food crops. Plant Pathologists have an important role to play in meeting the food security need of the nation by developing technologies for ensuring proper crop health. There is an urgent need to produce more food per unit area for feeding rapidly increasing human population. We need to act ahead of pathogens in our efforts to ensure normal crop yields. Cutting edge biotechnologies such as CRISPR gene editing systems and RNA interference are being exploited to rapidly developed crop varieties having resistance to specific pathogen races.

Crop protection has come a long way from predominantly chemical based approaches to more environment friendly IDM technologies including green chemistry. There is a need to develop bio-intensive modules for management of diseases. Decision support system need to be developed for important diseases for their timely, more efficiently management. There is need to adopt information technology at a larger scale to facilitate quick dissemination of new crop protection solutions among the farmers and other stakeholders.

Plant Pathology

The term Plant Pathology is derived from Greek words- *Pathos* (suffering) + *Logos* (study) i.e. the study of the suffering plant. Plant Pathology has two phases broadly:

- **Science-** *Understanding of the disease* i.e. the theoretical consideration of the suffering plants and *how do pathogens invade, how do plants defend themselves, what causes symptoms etc.*
- **Art-** *The application of the science to the field problems* i.e. Diagnosis and Control.

Plant pathology or phytopathology is the branch of agricultural, botanical or biological science which deals with the cause, etiology, resulting in losses and management methods of plant diseases.

Plant pathology can also be defined as the study of the nature, cause and prevention of plant diseases. Plant pathology is related to most of the old and new sciences like biology, physics, chemistry, physiology, mathematics, genetics, soil science, biochemistry, biotechnology etc.

Study of plant pathology includes the study of sciences *viz*, microbiology, bacteriology, virology, mycology, nematology, protozology, phycology, unfavorable , environmental factors, nutritional deficiencies and flowering plant parasites.

1. Microbiology: Study of microorganisms
2. Bacteriology: Study of bacteria
3. Virology: Study of viruses
4. Mycology: Study of fungi
5. Nematology: Study of nematode
6. Protozology: Study of protozoa
7. Phycology: Study of algae

Major Objectives

1. To study biotic, mesobiotic and abiotic causes of diseases or disorders.
2. To study the mechanisms of disease development by pathogens.

3. To study the plant (host)-pathogen interaction in relation to environment.
4. To develop methods of management of plant diseases.

Scope of Plant Pathology

1. Survey of crop fields to find out prevalence of diseases and their incidence.
2. Recording new diseases of economic importance if any, with their identification and extent of incidence.
3. Assessment of losses caused by different diseases of economic importance.
4. Study of etiology, symptoms, predisposing factors and recurrence of such diseases.
5. Find out suitable and economic methods of management of economically important plant diseases.
6. Assisting in breeding of disease resistant varieties.
7. To train the extension workers and subject matter specialists in order to bridge the gap between pathologists and farmers for better crop production.

Importance of Plant Pathology in Agriculture

1. Plant pathology has advanced the knowledge to protect the crop from losses due to diseases.
2. Most of the diseases with known disease cycle can now be avoided by the modification of cultural practices.
3. With the knowledge of mode of disease spread, many diseases of economic importance can now be managed .
4. Crop improvement and varietal resistance have been achieved against many diseases through the joint effort of breeder and plant pathologist.
5. The science of plant pathology has contributed disease free certified seed production.
6. Plant pathology has made possible to restrict the spread of plant diseases from one place to other and one country to other through suitable measures and quarantine legislation.
7. With the knowledge of plant pathology various prophylactic measures are adopted for successful management of diseases. These measures are seed treatment in seed borne diseases, soil treatment and crop rotations.
8. Diseases can be avoided in cold storage by the application of plant pathological measures as per recommendations made for different diseases for protection of fruits and vegetables in storage.

9. Plant pathology has made possible to recognize, discard or utilize toxic substances by producing toxic substances or by competition or by parasitism.

Plant Diseases

Disease is one of those terms that are very difficult to define. It is realized that disease (literally *dis-ease*) implies *lack of 'comfort'* and therefore, involves *deviation from normal functioning*. From time to time several definitions which have been proposed, in fact descriptive but not simultaneously exclusive. The definitions for the term disease are:

1. Disease is a malfunctioning process that is caused by continuous irritation, which results in some suffering producing symptoms.
2. Disease is an alteration in one or more of the ordered sequential series of physiological processes culminating in a loss of coordination of energy utilization in a plant as a result of the continuous irritation from the presence or absence of some factor or agent.
3. Disease is any morphological or physiological abnormality in a plant or any of its parts caused by continuous irritation and results in economic losses to human beings.
4. Any deviation from normal growth or structure of plants that is sufficiently pronounced and permanent to produce visible symptoms or to impair quality, and economic value.
5. A plant is said to be diseased when there is a harmful deviation from normal functioning of physiological process.

All these definitions indicate that disease

- Disease is not a pathogen but it is *caused by a pathogen.*
- It is not symptom but *results in symptoms.*
- It is not a condition as the condition results from disease and is not synonymous with it.
- It is *not an injury.*
- Disease cannot be infectious, it is actually the pathogen which is infectious.
- Disease results from continuous irritation.
- Disease is a malfunctioning process and this result in some suffering and therefore disease is a *pathological process*

Importance of Plant Diseases

Evidence indicates that plants were affected by disease 250 million years ago. The Bible and other early writings mention diseases, such as rusts, mildews, and blights, that have caused famine and other drastic changes in the economy of nations since the dawn of recorded history. Other plant disease outbreaks with similar far-reaching effects in more recent times are mentioned below:

S.No.	Major disease outbreaks	Country	Period
1.	Late blight of potato (Irish famine)	Ireland	1845–60
2.	Powdery and downy mildews	France	1851 and 1878
3.	Coffee rust	Ceylon	1870s
4.	Fusarium wilts of cotton and flax; southern bacterial wilt of tobacco	-	Early 1900s
5.	Sigatoka leaf spot and Panama disease of banana	Central America	1900–65
6.	Leaf spot of rice (Bengal famine)	India	1945
7.	Black stem rust of wheat	United State of America	1916, 1935, 1953–54
8.	Southern corn leaf blight	United State of America	1970
9.	Panama disease of banana	Asia, Australia, and Africa	1990 to present
10.	Coffee rust	Central and South America	1960, 2012 to present

Such outbreaks from plant diseases can have a significant economic impact, causing a reduction in income for crop producers and distributors and higher prices for consumers. Loss of crops from plant diseases may also result in hunger and starvation, especially in India where access to disease-control methods is limited and annual losses of 20 to 40 percent are not uncommon for major crops. The pointwire major importance of plant diseases are summerised as:

1. Plant diseases may limit the kinds of plants and industries in the area.
2. Plant diseases reduce the quality and quantity of plant produce.
3. Plant diseases may make plants poisonous to humans and animals.
4. Plant diseases may cause financial losses.
5. The cost of controlling plant diseases is also a direct loss to diseases.
6. Plant Disease also impact on enviroment sustainability, forest health recreational horticulture, human and animal health.

Questions

Long Answer

1. What is Plant Pathology? Discuss its importance in human welfare.
2. Define Plant Pathology. Write down the objective and scope of plant pathology in agriculture.
3. Plant pathologists proclaim that plant pathology is a discipline which encompasses both "art" and "science". What do you understand by the "art" of plant pathology?
4. What is disease? Mention importance of plant disease in agriculture with suitable classical examples.
5. Disdinguish between i.Science and art of plant pathology II.Disease and injury III. Disease and disorder IV. Soil invaders and Soil inhabitant.

Short Answer

1. Differentiate between the healthy (normal) and the diseased plant.
2. What are the objectives of Plant Pathology?
3. Define Plant Protection. Give its responsibilities to Agriculture?
4. Mention two important milestones in the development of the subject of Plant Pathology.
5. Which pathogens do not satisfy Koch's postulates and why?
6. Describe the socio-economic impacts of plant diseases.

2

History of Plant Pathology

History in general reveals chronological account of important events, contribution of persons who significantly influenced the thinking of their era and the interpretations of the observed facts or phenomenon over the period of time. The progress in plant pathology leading to the major land mark in mycology, plant bacteriology and plant virology will be discussed in chronological order for easy and better understanding.

Mycology

Year	Contributor (s)	Description
1500	Surapal	Wrote '*Vraksha Ayurveda*', the first book in which plant diseases were discussed
1665	Robert Hook	First observed Teliospore of *Pharagmidium disciflorum* under microscope
1676	Anton von Leeuwenhoek	Developed the *first microscope*
1729	PA Michaeli	Studied fungi and saw their spores on the pieces of water melon Wrote '*Nova Plantarum Genera*' Known as *Father of Mycology*
1743	John Needham	Reported plant parasitic nematodes in wheat galls
1755	M Tillet	Demonstrated that the *bunt of wheat was contagious* thought that the spores contained a poisonous entity.
1807	B Prevost	Proofed the *role of a micro organism in the causation of disease.* Demonstrated the control of wheat smut by steeping seed in copper sulphate solution
1821	EM Fries	Reported that smut and rust fungi as product of diseased plants Wrote " S*ystema Mycologium*" Known as *Linnaeus of Mycology*
1827	Cragie	Showed function of Puccinia in rust fungi
1840	LR Tulasne & C Tulasne	Confirmed the observation of Prevost with regard to the causal organism of wheat bunt. Known as *Reconstructor of Mycology*

1858	JG Kuhn	Published first text book on Plant Pathology namely "*The Diseases of Cultivated Crops: Their causes and Their Control*"
1861-65	Anton de Bary	First to indicate the nature of *obligate and facultative* forms
		Proved the real cause of late blight of potato is *Phytophthroa infestans*
		Described the sex and account of development in number of Phycomyetes and Ascomycetes
		Described the role of enzymes in tissue disintegration while working on soft rot of carrots caused by *Sclerotinia* spp.
		Demontrated *heterocious nature of stem rust of wheat*
1875	O Brefeld	Developed *pure culture technique*
1878	MS Woronin	Found out the life cycle of *potato wart* disease
1881	HM Ward	Reported the role of environment in the epidemiology of coffee rust
		Father of Tropical Plant Pathology
1882	Robert Hartig	Published a textbook -*Diseases of Trees*. He is called as "*Father of Forest Pathology*"
1885	PM Millardet	Discovered the *Bordeaux mixture* for the control of downey mildew of grapes
1885	Frank	Discovered *Mycorrhizal* fungi
1887	Mason	Introduced *Burgundy mixture*
1887	Jensen	Developed *hot water technique* for wheat smut
1894	Ericsson	Reported *Physiological specialization* in stem rust of wheat
1904	AF Blakeslee	Founded *heterothallism* in *Rhizopus*
1905	RH Biffen	Demonstrated inheritance of rust resistance in wheat in a Medelain fashion
1917	EC Stakman	Distinguished *biological forms* in cereal rusts
1923	Hansen and Smith	Term *Heterokaryosis*
1931	PA Saccardo	Wrote "*Syllogue Fungorum*"
1931	JC Luthra	*Solar heat treatment* for loose smut of wheat
1946	HH Flor	Gave *gene for gene hypothesis* while working on linseed rust
1952	G Pontecorvo and JA Roper	Discovered *parasexuality* in *Aspergillus nidulans*
1968	Vander Plank	Concept of horizontal and vertical resistance
		Father of Plant Epidemiology
1990	DF Klessig and I Raskin; JP Metraux and J Tyals	Demonstrated that *salicylic acid* is associated with systemic acquired resistance
2005	RA Dean and Co-workers	The first complete genome sequencing of a plant pathogenic fungus *Magnaporthe grisea*

Plant Bacteriology

Year	Contributor (s)	Description
1683	Anton von Leeuwenhoek	First observed bacteria
1876	Robert Koch	Proved that anthrax disease of cattle was caused by specific bacterium
1876	Louis Pasteur	Demonstrated role of bacteria in fermentation and decay
1876	Robert Koch	Described the theory called "*Koch's postulates*." He established the principles of *pure culture* technique
1876	Robert Koch and Pasteur	Disproved the theory of *spontaneous generation* of diseases and propose *germ theory* in relation to the diseases of man and animal
1876	Woronin	Isolated and described the *root nodule* bacteria in leguminous plant
1878	TJ Burril	Described first bacterial disease (Fire blight of apples) caused by *Erwinia amylovora*
1879	Prilleaux	Reported the bacterial decay of wheat kernels
1883	JH Wakker	Investigated yellow slime disease of hyacinth caused by bacterium
1887	EF Smith	Gave the final proof that bacteria could be the incitants of plant diseases
		First to notice and study *the crown gall disease*
		Father of Phytobacteriology
1980	DW Dye and Co-workers	Introduced *pathovar system* in taxonomy
1910	CO Jensen	Related *crown gall* of plants to cancer of animals.
1952	J Lederberg	Coined the term *plasmid*
1952	SA Waksman	Discovered *streptomycin*
1952	Zinder and J Lederberg	Discovered *transduction* in bacteria
1962	H Stolp	Discovered *bdellovibrios*
1972	PB New and A Kerr	Success in biological control of *A. radiobacter* strain K
1972	IM Windsor and LM Black	Observed a new kind of phloem inhabiting bacterium causing clover club leaf disease.
1972	Windsor and Black	Observed rickettsia like organisms in the phloem of clover plants infected with the club leaf disease
1974	I Zanen *et al.*	Demonstrated Ti plasmid in *Agrobacterium tumefaciens*.
1977	Chilton *et al.*	Showed that the crown gall bacterium transforms normal plant cells into tumor cells by introducing into them a plasmid, part of which becomes inserted into the plant cell chromosomes DNA

Plant Virology

Year	Contributor (s)	Description
1576	C Clusius	Recorded *Breaking of tulips* owing to viral infection
1886	AE Mayer	First to point out that *tobacco mosaic is readily transmissible and infectious* Demonstrated the sap transmission of the TMV
1892	D Ivanowski	Demonstrated that the agents of *TMV could pass through filters that retained bacterial cells* (bacteria proof filter)
1894	Hashimoto	Showed transmissiblelity of rice dwarf disease by leaf hopper (*Nephotettix apicalis var. cincticeps*).
1898	MW Beijernick	Concept of *contagium vivum fluidum* Father of Plant Virology
1917	d'Herelle	Discovered *bacteriophage*
1932	Knoll and Ruska	Invented *Electron microscope*
1935	WM Stanley	*Crystallized Tobacco mosaic virus*
1936	Bawden and NW Pirie	Found that the crystalline nature of the virus contains *nucleic acid and protein*
1939	Kausche, Pfankuch and Ruska	Showed virus particles for the first time with the electron microscope Confirmed that TMV was rod shaped
1942 & 1944	Muller Williams and Wycoff	Developed *shadow casting technique* with heavy metals which was useful for determining the overall size and shape of the virus particles
1949	Markham and Smith	Isolated TYMV and showed that it contained (1) An infectious nucleoprotein (about 25%) and (2) non infectious protein
1952	Morel and Martin	Showed that virus free plants could be obtained from totally infected parents using *meristem tip culture*
1954	Kassanis	Showed that virus could be eradicated from infected plants by high temperature treatment
1956	Gierer and Schramm	Proved that the *nucleic acid fraction* of the virus is actually the infectious agent
1966	Kassanis	Discovered the *satellite viruses*
1963 & 1966	Black and Markhan Miura *et al.*	Showed that wound tumour and rice dwarf viruses contain double started RNA.
1967	Doi and Co-workers	Discovered *Mycoplasma like organism*
1967	Ishiie *et al.*	Showed that the MLO bodies and the symptoms disappeared temporarily when the plants were treated with *tetracycline antibiotic*
1967	Diener and Raymer	Discovered *viroid*
1968	Shephard *et al.*	Showed that cauliflower mosaic virus contains *double stranded DNA*
1974	Davies and Worley	Coined the term *spiroplasma*

1977	Clark and Adams	Develpoed *ELISA assay* for detection of plant viruses
1981	TW Randles and co-workers	Discovered *virusoids*
1982	Prusiner	Discovered *Prions*
1994	Sears and Krikpatrick	MLO that infects plants have been reclassified as Phytoplasmas

Plant Disease Control

Year	Contributor (s)	Description
1000	Homer	Mentioned about the use of *sulphur* in plant disease control
1807	B Prevost	Recommended copper sulphate for wheat seed treatment against bunt and demonstrated *first time the fungitoxic value of copper compounds*
1882	PMA Millardet	Discovered *Bordeaux mixture* for control of downy mildew of grape vine
1887	Mason	Discovered *burgundy mixture*
1921	Bewley	Developed *chestnut compound*
1934	Tisdale and Williams	Reported fungitoxicity of *dithiocarbamates*
1942	Singh	Delveloped *Chaubatia* paste
1943	SA Waksman and A Schatx	Discovered *Streptomycin*
1952	Kittleson	Discovered *captan* also known as Kittleson's Killer
1966	Von Schmeling and M Kulka	Discovered *Systemic fungicides* oxanthin

Plant Pathology in India

Year	Contributor (s)	Description
1885	KR Kritikar	First Indian scientist who collected and identified fungi
1886-1971	JF Dastur	First Indian Plant pathologist known for the establishment of *genus Phytophthora* and diseases caused by it in castor and potato
	TS Sadasivan	Developed concept *of vivotoxins* and worked out the mechanism of wilting in cotton owing to *Fusarium oxysporum f.sp. vasinfectum*
	SN Dasgupta	Carried out exensive study *on black tip* of mango
	MJ Thriumalachar	Discovered several antifungal antibiotics *viz* aureofungin, haymycin etc.
	YL Nene	Reported *Khaira disease* of rice caused due to *Zn deficiency* Authored book "*Fungicides in Plant Disease Control*"
1889	DD Cunningham	Reported the causal organism of *red rust of tea* in Assam caused by *Cephaleurous virescens*
1920	EJ Butler	First direcotor of Imperial Mycological Institue in England Initiated an exhaustive study on Indian fungi and the disease caused by them

		Discovered genus Allomyces Wrote "*Fungi and Diseases in Plants*" *Father of Indian Plant Pathology*
1940	KC Mehta	Studied epidemiology of cereal rusts in India Write monograph on "*Further studies on cereal rust in India*" *Father of Indian Rust*
1948	BB Mundukar	Established Indian *Phytopathological society* with its journal "*Indian Phytopathology*" Wrote Fungi and Plant Diseases
1953	JC Luthra and A Sattar	Developed *solar heat treatment* for loose smut of wheat
	Ramanujam and Co-workers	Developed *seed plot technique* for virus free seed potato production in Indo Gangetic plains
1975	S Nagrajan and H Singh	Formulated Indian Stem Rust Rules for *Puccinia graminis tritici*

Questions

Long Answer

1. Write an essay on developments of various disciplines of plant pathology upto 19th century.
2. Give a concise account of the main Indian contributions in plant pathology.
3. Write notes on:

a) Discovery of Bordeaux mixture.

b) Adolf Mayer's contribution in development of virology.

c) Germ theory of disease.

d) Multidisciplinary nature of plant pathology.

4. Mention the most important contributions of the following Indian Plant Pathologists.

i) KC Mehta

ii) BB Mundkar

iii) MJ Thirumalachar

iv) GS Kulkarni

5. Name five contemporary scientists of Plant Pathology in India.

Short Answer

1. Write an essay on history of Plant Pathology ?
2. Name two ancient Indian holy books in which reference to plant diseases and their control devices is made.

3. What would have been the main reason, in your opinion, behind they very slow growth of plant pathology upto the time of Antony Van Leeuwenhoek?
4. Who is credited to have laid the foundation of the "science" of plant pathology? How?
5. Name the authors of the books "Fungi and Diseases in Plants" and "Fungi and Plant Diseases".
6. What were the conclusions made by Ivanowski (1882) and Beijerinck (1898) in reference to the experiments on tobacco mosaic disease?

3

Terms and Concepts in Plant Pathology

There is a very impressive and extensive number of terms and definitions used in plant pathology, many of which you will not come in contact with. With interest, study and practice, terms and names will come to light and become familiar to you. When you start out, don't worry too much about the terms; but also don't be afraid of them. The more exposure you have to the subject the more comfortable you will be when dealing with your peers and the public. With these you are going to be exposed to some terms and definitions that will be a *good ice breaker* and prepares you for the information to follow:

Terms

- **Plant pathology** is the study of plant diseases and the abnormal conditions that constitute plant disorders.
- **Etiology** is the determination and study of the cause of disease.
- **Pathogen** is an organism which causes a disease.
- **Pathogenicity** is the capability of a pathogen to cause a disease
- **Plant disease** is an abnormality in the structure and/or function of the host plant cells and/or tissue as a result of a continuous irritation caused by a pathogenic agent or an environmental factor.
- **Plant injury** is an abrupt alteration of form or function caused by a discontinuous irritant. Plant injury includes insect, animal, physical, chemical or environmental agents.
- P**arasite** is an organism which lives on or in another organism and obtains its nutrition there from.
- **Obligate parasite** is an organism which is wholly dependent for its nutrition on another living entity. Obligate parasites are **biotrophs** which also depend entirely on a living host for its nutrition.
- **Autotroph** is a plant that can make its own food through photosynthesis.
- **Facultative parasites** are organisms that gain their nourishment by digesting dead organic material.

- **Host** is an organism (eg. a plant) that is harboring a parasite or pathogen from which it obtains its nutrients.
- **Resistant**: Ability to exclude, hinder or overcome the effects of a given pathogen or other damaging factor.
- **Tolerance** is the ability of a plant to be colonized by a pathogen or exposed to an abiotic factor without dying or demonstrating disease symptoms.
- **Susceptibility** is the antithesis of resistance. Inability of plant to overcome any kind of disease.
- **Signs** are the physical evidence of damage caused by biotic or abiotic agents such as the pathogen itself, pests, spores, fruiting bodies, bacterial ooze and so forth.
- **Symptoms** are the visible response of a plant to biotic and/or abiotic factors that result in a change or abnormality in the plant.
- **Syndrome** is the totality of the effects demonstrated in a host by one disease, whether simultaneously or successively, and whether visible to the unaided eye or not.
- **Diagnostics** is the determination of the nature and/or cause of a disease or disordered condition.
- **Life cycle** is the sequence of distinct events, such as sexual reproduction, that occur between the appearance and reappearance of the causal organism.
- **Inoculation** is the process of applying inoculum to a host.
- **Polycyclic** disease is one that completes two or more life cycles in one year.
- **Monocyclic** disease is one that has one life cycle in one year.
- **Transmission** usually implies active transfer by means of grafting, insects, mechanical factors, animals and so on. To disseminate or spread means to disperse or distribute.
- **Disseminate** usually refers to long- distance distribution.
- **Spread:** Refers to local distribution.
- **Vectors** are active agents of transmission such as insects, mites, nematodes and other animals.
- **Infection** is the establishment of a parasite on or within a host cell or tissue.
- **Colonization** of a host results from the establishment, growth and reproduction of the pathogen on or in infected plant.

- **Infestation** refers to the establishment on the surface of a host by a large number of insects or other animal pests. With infestation there is no implication that infection has occurred.
- **Epidemic** is the unarrested, widespread increase of an infectious disease, usually limited in time.
- **Endemic** disease is one that is permanently established in a moderate or severe form within a defined area. Endemic diseases usually become indigenous following initial introduction of the pathogen.
- **Epidemiology** is the study of factors affecting the outbreak and spread of infectious disease.
- **Taxonomic classification** is the systematic ordering of plants and animals.

Concept

The concept that plant disease is not caused by a single organism or disease agent is overlooked by most people. Plant disease is instead caused by a combination of three factors: a pathogen, environmental conditions and host response. The *concept* is visualize by the *disease triangle* in which the three factors , *pathogen*, *environment* and *host* make up the three sides of a triangle and the area of triangle is the amount of disease produced in the plant. The disease triangle shows visually that a very weak and inefficient pathogen could cause substantial disease if the environment is conducive or if the host is extremely susceptible.

Questions

1. Define: Disease, Pathogen, Pathogenicity, Pathogenesis, Virulence, Host, Predisposition, Inoculum potential, Disease potential, Disease Severity, Inoculum density.
2. Distinguish between a. Pathogen and Parasite b. Pathogenicity and Virulence c. Monocyclic and Polycyclic disease d. Virulence and Aggressivenes.
3. Write notes on i. Disease triangle ii. Disease tetrahedron iii. Disease prevalence iv. Disease incidence v. Competitive saprophytic ability.
4. What are the component of soil environment? Explain how do they influence the disease.
5. What are the component of disease triangle? Explain the role of aerial environment.

4

Pathogenesis

Definition

Pathogenesis can be defined as the chain of events leading to the development of a disease. In another words it is also called *disease cycle.*

It tells us about

- Source of perennation of the pathogen.
- Mode of its dispersal from the source of survival and during its spread of disease.
- Help us to formulate effective control measures.
- The events constituting pathogenesis occurs in several well defined steps, one after the other, to complete the disease cycle.
- Pathogen perennates at some location during the absence of cultivated host.
- Its gets transported to the cultivated host through some agencies.
- The pathogen breaks the host barrier to establish infection.
- Effect the host physiology, damage the plant and the plant expresses symptoms.
- Finally the pathogen finds an exit from the host through its propagules.

Stages in the Pathogenesis or Disease cycle

The primary events in a disease cycle are

1. Inoculation
2. Penetration
3. Establishment of infection
4. Colonization (Invasion)
5. Growth and Reproduction of the pathogen
6. Dissemination of the pathogen
7. Survival of the pathogen

1. Inoculation

Inoculation is the coming in contact of a pathogen with a plant. Inoculum is defined as the part of the pathogen, which on contact with a suitable host can cause infection. E.g. of fungal inculum; chlaymydospores, sclerotia, oospores, teleutospores, zoospores, conidiospores etc. In bacteria, mollicutes, protozoa virus, viroids the inoculum is whole individual. In nematodes, the inoculum may be adult nematodes, nematode juveniles, or egg. One unit of inoculum of any pathogen is called propagule.

a) Types of Inoculum

All inoculum that survives in the winter or summer and causes the original infection in the spring or autumn is called *primary inoculum*, and the infection it causes are called primary infections. An inoculum produced from primary infections is called *secondary inoculum* and it in turn causes secondary infection.

b) Source of Inoculums

1. Sowing and planting materials *viz*., seeds, tubers, bulbs, rhizomes.
2. Diseased plants in the field.
3. Debris left on the land or used in compost heaps.
4. Host plants in a dormant stage when conditions are not favourable.
5. Soil infected with certain soil borne pathogens.
4. Spread of plant diseases.

Plant diseases are spread in various ways:

1. Spread through seeds and other planting materials such as tubers, rhizomes etc.
2. Spread by natural agents such as wind, water and rain.
3. Spread by animal and insects.

2. Penetration

The penetration process has two phases:

i) Pre penetration stage

ii) Penetration

i) Pre penetration stage

During the pre penetration stage the fungus hyphae or spores come in contact with the host surface. The spores will germinate to produce germ tubes or infection threads. This stage may be abortive if environments do not favor growth of the hypha or germination of spores.

ii) Penetration

During penetration, the infection thread enters the host by any of the following methods.

a) Entry through natural openings like stomata & lenticels.

b) Entry through rupture of the host surface due to development of organs like prop roots.

c) Entry through wounds due to mechanical injury or insect injury.

d) Direct penetration:'

- Mechanical pressure by the infection thread. The pathogen exerts its own effort to break the host barrier and directly enter through cuticle or epidermis without seeking the wounds or natural openings. Host barriers can be either structural barriers or chemical barriers.
- Chemical action : enzymes produced by the pathogen will break down the epidermal cells and thus gain entry.

Pathogens penetrate plant surfaces by direct penetration. Some fungi penetrate tissues in one way only, others bacteria enter plant mostly through wounds, less frequently through natural opening and never directly. Viruces, viroids, mollicutes and protozoa enter through wound made by vectors, although some viruses and viroids may also enter through wounds made by tools and other means. Nematodes enter plants by direct penetration and sometimes through natural openings. Penetration does not always lead to infection.

3. Infection

Infection is the process by which pathogen establish contact with the susceptible cells or tissues of the host and take nutrient from them. During infection pathogens grow or multiply or both within the plant tissues and invade and colonize the plant to a lesser or greater extent. Therefore, invasion of the plant tissues by the pathogen and growth and reproduction of the pathogen (colonization) in or on infected tissues are actual two concurrent substages of disease development within the stage of infection. Successful infections result in the appearance of symptom on the host plant. In most plant diseases, symptoms appear from a few days to few weeks after inocolation.

4. Colonization

Various pathogen invade hosts in different ways and to different extents. Most fungi spread into all the tissues of the plant organs they infect, either by growing directly through the cells as an intracellular mycelium or by growing between the cells as an intercellular mycelium. The fungi that cause vascular wilts invade the xylem vessels of plants. Bacteria invade tissues intercellularly, although when part of the cell wall dissolve, bacteria also grow intracellular. Bacteria causing vascular wilt, invade the xylem vessels. Viruses, viroids, mollicutes, fastidious bacteria and protozoa invade tissues by moving from cell to cell intercelluarly. Viruses and viroids invade all types of living plant cells, mollicutes and protozoa invade phloem sieve tubes and perhaps a few adjacent phloem parenchymatous cells. Most nematodes invade tissues intercellularly, but some can invade intracellularly-many nematodes do not invade cells or tissues at all but feed by piercing epidermal cells with their styles.

5. Growth and reproduction of the pathogen

Individual fungi and parasitic higher plants generally invade and infect tissues by growing into them from one initial point of inoculation. Most of these pathogens, whether producing a small spot, a large infected area, or a general necrosis of the plant continue to grow and branch out within the infected host indefinitely so that the same pathogen individual spread into more and more plant tissues until the spread at the infection is stopped or the plant is dead. All other pathogens like bacteria, virus, viroids, nematodes and protozoa, do not increase much, if at all, in size with time, since, their size and shape remain relatively unchanged throghout their existance. These pathogens invade and infect new tissues within the plant by reproducing at a rapid rate and increasing their number tremendously in the infected tissues. The progeny may then be carried passively through:

a) Plasmadesmata: virus and viroids

b) Phloem: viruses, viroids, mollicutes protozoa, some fastidious bacteria

c) Xylem: some bacteria nematodes

6. Dissemination of the pathogen

Following the infection, the pathogen will continue its growth, produce spores or reproductive units, which will find exit through host surface and spread to repeat the same process. Many pathogens spread through crops in a special and often spectacular way. Some cover tremendous distances at a remarkable speed, especially pathogens with airborne propagules, like rusts. Pathogens which infect subterranean plant organs exhibit a restricted pattern of dispersal. Then there are pathogens which depend for their dispersal of spores and propagules on rain or water, while others need a vector like man , insects and nematodes. Some parasites are transported with seed, thus depending on the hosts own dispersal processes.

There are many dispersal mechanisms. Some of these mechanisms are both elegant and highly efficient in that they are closely adapted to the biology of the host, like the synchronization of spore release in certain pathogens like Venturia and Claviceps. Other parasites produce millions of spores, saturating the environment with propagules in an apparently haphazard and wasteful way.

7. Survival of Pathogen

Pathogens must find some alternate source of their survival in the absence of their cultivated host; otherwise the infection chain will be remain incomplete. Infection chain is the chain of events leading to the completion of pathogenesis.

The source of survival can be grouped into

a) Infected host as reservoir of inoculums

- Cultivated host : Main host
- Collateral host : Wild host of the same family
- Alterante host: Wild host of other family

b) Saprophytic survival outside the host. Soil and plant debris serve as the media for survival. E.g. Pythium, Rhizoctonia

c) Dormant organs of pathogen as a source of survival and primary inoculums; only fungi and nematodes have spores and cysts etc, while virus and bacteria have no resting stage.

Questions

1. Explain the following: i. Infection ii. Latent infection iii. Invasion iv. Colonization v. Inoculum potential
2. Describe the general principles of plant infection? How infection is caused in some of the important plant pathogen.
3. What is infection? Describe in detail the mechanism of infection and establishment of any typical fungal pathogen.
4. What are the different stages in the development of a infectious plant disease. Describe in detail the pre penetration activities of fungal pathogens. Gives suitable examples.
5. Describe various aspects of a typical disease-cycle.

6. Write notes on:

a) Polycyclic diseases

b) Air-borne diseases

c) Koch's postulates

d) Irish famine

e) Disease manifestation

7. Discuss in detail the prepenetration phase of infection.

8. Write notes on:

a) Chemotaxis

b) Melanized appressoria

c) Pathogen adhesion on host

d) Infection plaques

9. Discuss the growth of pathogen superficially, subcuticularly, and endophytically in their hosts.

5

Factors Affecting Disease Development Disease Triangle and Tetrahedron

The Disease Triangle and Disease Pyramid

The disease triangle is one of the first concepts encountered by students in an introductory plant pathology course and often may be encountered in higher level classes. The interactions of the three component namely (1) *susceptible host* (2) *virulent pathogen* and (3) *favorable environments* have been visualized as a triangle, generally referred to as the "*DISEASE TRAINGLE*". Each side of the traingle represent one of the three components. The length of each side is proportional to the sum total of the characteristic of each component that favor disease. For example, the more virulent and abundant the pathogen, the longer the pathogen side would be and the greater the potential amount of disease. Similarly if the plant are resistant, the widely spaced or wrong age, the amount of disease would be small or zero. Where as if the plants are susceptible, at a susceptible stage of growth, or densely planted, the host side would be long and the potential amount of disease could be great. Also, the more favourable the environmental conditions that help the pathogen, or that reduce host resistance, that longer the envirnmental side would be and greater the potential amount of diseases. If the three component of the disease traingle could be quantified, the area of the traingle would represent the amount of disease in a plant or in plant population. If any of three component in zero, these can be no disease.

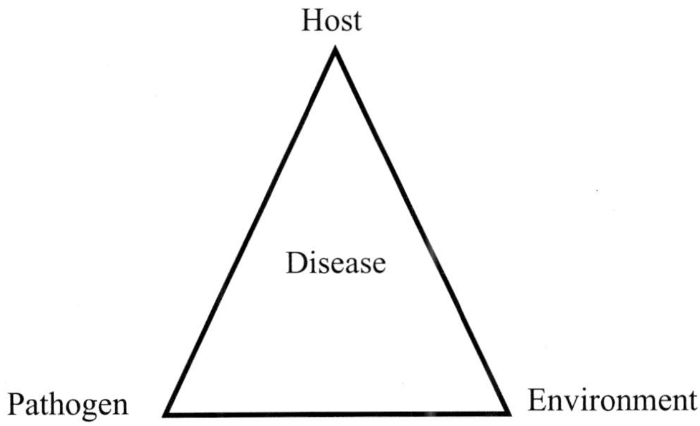

Some plant pathologists have elaborated on the Disease Triangle by adding one or more parameters. Suggested additional parameters may be *human activity, vectors* and *time.* A three dimensional disease triangle may result in four dimensional *DISEASE PYRAMID* or *DISEASE TETRAHEDRON* after addition of a single parameter.

Factors affecting disease development

There are three major factors that affect disease development

1. Pathogen factors
2. Host factors
3. Environmental factors

1. Pathogen factors

- Presence of pathogen
- Pathogenicity (virulence and aggressiveness)
- Adaptability
- Dispersal efficiency
- Survival efficiency
- Reproductive fitness

2. Host factors

- Susceptibility
- Growth stage and form
- Population density and structure
- General health

3. Environmental factors

- Temperature
- Rainfall
- Dew
- Leaf wetness period
- Soil temperature
- Soil moisture
- Soil fertility
- Air pollution

1. Pathogen factors

a) **Presence of pathogen:** The main factor which determines whether or no disease occurs at all is the presence or absence of the pathogen. In the out break of late blight of potato in north India, the presence of absence of the pathogen is the overriding factor that determines whether or not disease occurs.

b) **Pathogenicty:** The amount of disease that develops if often determined by pathogenicity of the prevalent population of the pathogen. The term pathogenicity comprises both the virulence of the pathogen and its aggresivenenss. It also includes aspects of dispersal and survival fitness. Control of diseases often involves monitoring the pathogenicity of the prevailing pathogen population as a guide to the breeding of resistant crop varieties.

c) **Adaptibility:** The adaptability of pathogen is very important in determining its ability to overcome resistance in newly released varieties or to adapt to the changed environmental conditions. It is governed by the genetic flexibility of the pathogen population and its reproductive efficiency. *Puccinia graminis* has high reproductive efficiency and so is likely to adapt very fast where as *O. theobromae* on cocoa has a much lower reproductive capacity and so likely to adapt more slowly.

d) **Dispersal efficiency:** The ability of a pathogen to cause destructive disease epidemics depends on its ability to disperse rapidly over long distances. The urediniospores of rust can be blown long distances in a few days , allowing epidemics to develop rapidly over very large areas. Soil borne pathogens, on the other hand, have limited dispersal ability and ted to cause localized disease outbreaks.

e) **Survival efficiency:** The ability of a pathogen to cause disease in successive seasons depends on its ability to survive. *Plasmodiophora brassicae*, the cause of club roots of crucifers, form spores that enable the pathogens to survive in the soil for many years, thus reducing the effectiveness of disease control using crop rotation. The sclerotia of *Sclerotium rolfsii* and *Sclerotinia* spp. are also particularly effective in enabling these pathogens to survive through the summer and winter, resulting in larger quantities of inoculums to initiate epidemics in the new growing season.

2. Host factors

a) **Susceptibility:** The occurrence of individuals in the host population that are *susceptible* to the particular pathogen is the main factor affecting disease development. For a disease epidemic to occur, the host plant population must be largely susceptible to attack by the pathotypes of the pathogen in the vicinity. The most important means of disease management is to plant varieties that are not susceptible to the prevailing population of the pathogen of concern.

b) **Growth stage and form:** The growth stage and form of the host greatly influence the occurrence of disease. The growth stage also determines the degree of closure of the canopy, which in turn affect the microclimate within the canopy. Some diseases such as damping off are more common in seedlings, while others are characteristic of mature plants.

c) **Population density and structure:** The characteristics of population structure and density in the host will have a large bearing on the development of disease in the population. Many plant diseases are much more common in dense than in sparse plantings. Modern intensive agriculture provide and ideal situation for development of plant disease because variety or even clones derived from a single plant are commonly planted over a large area with few or no individuals of non host species or cultivars interfere with inoculums dispersal. Spectecular disease epidemics have occurred when susceptible host varieties were planted over large areas or when a new pathogen was introduced such plantations. Such a situation occurred with the spread of late blight of potato to Ireland in the 1840s and with the spread of coffee rust to Ceylon in the 1870s. Both pathogens encountered extensive monocultures of their respective hosts, potato and coffee and caused destructive epidemics.

d) **General health:** The prior health of the host is often important in determining the occurrence of disease. Necrotrophic pathogens are often more damaging on poorly growing than on vigorous hosts while opposite is the cause of biotrophs.

3. Environmental factors

The total environment encompasses the whole biosphere within which the epidemiologic processes join up. The *total environment* is divided into *micro, meso* and *macro-environemnt* for easy understanding.

The *microenvironment* is the space in which the epidemiologic processes at cell and organ level occur. The largest part of the infection cycle, from *deposition to takeoff* takes place in microenvironment. The microenvironment of the leaves is called *phyllosphere*, which includes a laminar air layer of *up to 1mm in thickness* surrounding the leaf. The *phyllosphere* is a *three dimensional space*, whereas the actual leaf surface, or *phylloplane*, is a *two dimensional space*. In regard to the roots there is a *three dimensional rhizosphere* (Hiltner, 1904) & a two dimensional *rhizoplane.*

The mesoenvironment is formed by the crop. It is formed by the plants and simultaneously it influences the plants and the epidemiologic processes in the crop.

The macroenvironment is the air layer from the crop surface to the troposphere. Some epidemiologic processes of extreme importance occur in the macroenvironment. There are three major environemental factors that affect disease development are.

a) **Temperature:** Temperature affects *Latent period*: the time between infection and the first appearance of disease symptoms, *Generation time*: the time between infection and sporulation and *Infectious period*: the time during which the pathogen continues producing propagules. At higher temperatures the disease cycle is speed up with the result that epidemics develop faster. Under cooler conditions, epidemics progress is usually slower. So disease incidence and severity may not reach the threshold level necessary to cause significant crop loss.

b) **Moisture:** Moisture is the most important environmental factor influencing disease outbreak caused by fungi, bacteria and nematode. The term "moisture" cover *rainfall, relative humidity, dew* and *leaf wetness.* Moisture has an important effect on dispersal of pathogens. Free water or the impact of raindrops facilitates the liberation and dispersal of many fungi and nearly all bacteria. This is very useful adaptation for a pathogen because when the propagules are dispersed, conditions are also likely to be suitable for germination and infection. For examples coelomycetes produce wet spores that are picked up by rain splash and dispersed. Many dry spores are liberated by the force of impact of raindrops. The influence of rain splash and running water on dispersal of pathogen propagules is also important. Leaf wetness is a more accurate predictor of disease than rainfall and dew period. The combination of leaf wetness and ambient temperature is critical in determining the proportion of pathogen propagules that infect the host.

Soil borne diseases are affected mainly by the environmental conditions in the soil. Wetness is often a critical factor. Some pathogens like pythiaceous fungi causing damping off require a period of soil saturation to allow germination of survival prpagules and infection of roots. The zoospores of pyhecious fungi require free water for their release and mobility.

c) **Soil fertility**: Soil fertility can affect development of both soil and seed borne disease. Some facultative pathogens cause more disease when plants are growing poorly under conditions of nutrient deficiency. On the other hand biotrophic pathogens such as rusts and powdery mildews are often more common on vigorous well fertilized plants than on poorly growing plants.

The organic matter content of soil has an important influence on some soil borne pathogens. For example, eucalyptus dieback is common in sandy or gravelly soils low in organic matter. Successful management of *P. cinnamomi* in avocado has been achieved by greatly increasing the organic matter content. Soils high in organic matter contain larger populations of antagonistic microorganisms which reduce the survival time of plant pathogens and the subsequent incidence of disease.

d) **Pollution:** The aerial environment has been observed to affect plant disease. In regions where aerial pollution associated with industrial development has been especially great, the concentration of pollutant like oxide of sulphur and nitrogen can affect disease development and in the form of acid rain, can directly damage plants.

Questions

1. Name four parasitic angiosperms that causes diseases on angiosperms
2. What is an epidemic?
3. Which of the following two statements are correct and why?
   a) Disease cycle is an alternate name of the life cycle of a pathogen.
   b) Disease cycle represents the life cycle of the disease, not the life cycle of the pathogen that causes the disease.
4. What is systemic infection?
5. How root-cap border cells are related to the infection in soil-borne diseases?
6. Name any two attractants stimulants of soil-borne pathogens.
7. Which bacterial pathogen is called 'natural genetic engineer' and why?
8. What is 'infection cushion'?
9. Which category of pathogens develop haustoria? Mention possible functions of haustoria apart from nutrient uptake.

6

Classification of Plant Diseases

There are number of diseases which attacks crop plants. Therefore It is mandatory to classify plant disease into more manageable and more specific group based on several criteria for easy and better understanding. Let us discuss one by one.

1. On the basis of type of infection

On the basis of type of infection disease can be classified into two types

i) **Localized disease-** Affecting only a part of the plant; leaf spots and anthracnoses

ii) **Systemic disease-** Affecting the entire plant like mosaic and leaf curl disease.

2. Based on their occurrence

Based on their occurrence the disease can be divided into four groups:

1. **Endemic disease**: When a disease is more or less constantly present in one form or other or less constantly present from year to year in a moderate to severe form, in a particular country or part of earth, it is classified as endemic. Ex. Wart disease of potato in Darjelling districts

2. **Epidemic or epiphytotic diseases**: The term 'epidemic' means '*among the people*' and in true sense applies to those diseases of human beings which appear very virulently among large section of the population. To carry the same sense in the case of plant diseases, the term ***epiphytotic*** has been coined. An *epiphytotic* disease is one which occurs *widely but periodically*. It may be present constantly in the locality but assumes severe form only on occasions. Ex. Irish famine due to late blight of potato

3. **Sporadic diseases**: Sporadic diseases are those diseases which occur at very irregular intervals and locations and in relatively few instances. Ex. Potato spindle tuber.

4. **Pandemic diseases**: When a disease is prevalent throughout the country, continent or the world it is known as a *pandemic* disease. Latest example corona disease.

3. Based on the cause

Based on the cause the disease can be divided into two parts

1. **Infectious or Biotic disease**:These are diseases incited by biotic or mesobiotic agents under a set of suitable environments. Example, fungal disease, bacterial disease, viral disease.
2. **Non-infectious or Abiotic disease**: These are the diseases with no biotic or mesobiotic agents associated, remain noninfectious and cannot be transmitted from one diseased plant to another healthy plant. Example. Disease caused by nature-frost, rain, wind, sun, hail storm.

4. Based on the production of inoculum

Based on the production of inoculum, the disease can also be divided into two parts;

1. **Single cycle disease (Simple Interest Disease):** When the increase of disease is mathematically analogues to simple interest of money, it is called simple interest disease. There is only one generation of disease in the course of one epidemic. Such diseases develop from a common source of inoculums i.e the capital i.e inoculum is constant. Example; Loose smut of wheat.
2. **Multiple cycle disease (Compound Interest Disease):** When the increase in disease is mathematically analogues to compound interest of money, the disease is called compound interest disease. There are several or many generations of the pathogen in one life cycle of the crop. The capital ie inoculum is increased by the amount of interest. Example; Late blight of potato.

5. On the basis of type of perpetuation and spread

On the basis of type of perpetuation and spread, the disease can be divided into three groups.

1. **Soil-borne diseases**: The causal agents perpetuate and spread through soil. Example. Damping off and Fusarium wilt.
2. **Seed-borne diseases**: Seed or seed materials help in the perpetuation and spread of this disease. The disease causing agents may be internally seed-borne or externally seed-borne. Example: Loose smut of wheat (Example of internally seed borne), Blast of rice (Example of externally seed-borne).

3. **Air-borne diseases**: In these type of diseases the causal agents are spread by wind (air). Example. Early leaf spot and late leaf spot of groundnut.

6. Based on symptoms

Based on symptoms the disease can be classified into rusts smuts, wilts, blights, cankers, mildews, rots, damping-off, die-back, scab etc.

7. Based on the host plant

On the basis of host diseases are classified like diseases of apple, diseases of wheat, diseases of rose, diseases of coconut, diseases of coffee, diseases of cotton.

Questions

1. What is a plant disease? How will you classify diseases of plants on the basis of their occurrence.
2. What are the causes of diseases in plants? Give the general classification of plant diseases.
3. Give the definitions and examples of the following types of diseases.

i) Endemic diseases

ii) Epiphytotic diseases

iii) Sporadic diseases

iv) Pandemic diseases.

4. Classify plant diseases based on production of inoculums.
5. Distinguish between bioatic and abiotic pathogens.

7

Important Plant Pathogenic Organisms

Plant diseases are classified on the basis of type of pathogenic or non-pathogenic causes of the disease. The classification is based on the plant pathogenic organisms as follows.

A. Parasites

The diseases are incited by parasites under a set of suitable environment. Association of definite pathogen is essential with each disease. They include both biotic and mesobiotic agents.

i) **Biotic agents:** They are also called as animate causes. They are *living organisms*.

Biotic agents include

1. Prokaryotes

a) True bacteria or bacteria (Facultative parasites) e.g. Citrus canker.

b) Rickettsia-like bacteria (RLB) e.g. Citrus greening, Pierce's disease of grape

c) Mollicutes or wall-less prokaryotes.

i) Mycoplasma-like organism (MLO) e.g. Sesame phyllody, Little leaf of egg plant.

ii) Spiroplasma e.g. Corn stunt, Citrus stubborn.

2. Eukaryotes

a) **Protists** (Unicellular, coenocytic or multicellular with little or no differentiation of cells and tissues).

i) Fungi e.g. wilt of cotton.

ii) Protozoa e.g. heart rot of coconut.

iii) Algae e.g. red rust of mango.

b) **Plants** - Parasitic flowering plants or Phanerogamic parasites - Broomrape of tobacco etc.

c) **Animals**: Nematodes-Root knot nematode etc.

ii) **Mesobiotic agents**: They are *infectious agents*. They can be crystallized and are considered non-living. But their multiplication in the living plants ensures that they are living. Hence they are called as mesobiotic agents. They include viruses and viroids.

a) Viruses e.g. yellow mosaic of blackgram

b) Viroids e.g. spindle tuber of potato

B. Non-parasites or Abiotic agents

They are also called as *non-infectious or physiological disorders*. When no pathogen is found, cultured from or transmitted from a diseased plant, then the disease is said to be caused by a non-living or environmental factor. These diseases occur because of disturbances in the plant system by the improper environmental conditions in the air or soil or by mechanical influences. They are listed below.

i) Too low or too high temperature

ii) Lack or excess of soil moisture

iii) Lack or excess of light

iv) Lack of oxygen

v) Air pollution

vi) Mineral deficiencies or toxicities

vii) Soil acidity or alkalinity

viii) Toxicity of pesticides

ix) Improper agricultural practices.

Questions

1. What are the biotic, abiotic and mesobiotic pathogens of plant diseases? Give in brief the classification of plant diseases according to major causal agents.
2. Disdinguish between biotic and abiotic pathogens .
3. Give two examples each of biotic, abiotic and mesobiotic pathogens.
4. Discuss about physiological disorders in plants.
5. Differentiate between pathogen and parasites.

8

Different Groups Fungi, Bacteria, Fastidious Vascular Bacteria, Phytoplasmas, Spiroplasmas, Viruses, Viroids, Algae, Protozoa, Phanerogamic Parasites and Nematodes with Examples of Diseases Caused by Them

Fungi

Definition

Fungi are eukaryotic, spore bearing, *achlorophyllus, heterotrophic* and thallophytic plants with varied forms and habitats, representing heterothallic or homothallic types of sexuality with characteristic sexual and /or asexual means of reproduction without tissue differentiation.

Main Characteristic for Fungi

The main characteristics of fungi are as follows:

Distribution	Cosmopolitan
Thallus	Unicellual, filamentous(mycelia), or occasionally plasmodial
Cell wall	Made up of chitin (Cellulose only in class Oomycetes)
Chlorophyll	Lack chlorophyll
Nucleus	Eukaryotic
Life cycle	Simple to complex
Sexuality	Asexual and sexual
Habitat	Ubiquitous, saprobes, parasites or symbionts
Nutrition	Heterotrophic and absorptive
Storage product	Oil and glycogen

Major diseases caused by fungi

i) Late blight of potato

ii) Powdery mildew of pea

iii) Downy mildew of pea

iv) Anthracnose of mango

v) Black stem rust of wheat

vi) Blast of rice

Bacteria

Definition

Bacteria belong to proakaryota which encompasses organisms with a primitive type of nucleus lacking a clearly defined membrane .The bacteria are smaller than fungi and measure about 0.5 to 1.0 x 2.0 to 5.0µ.

Main characteristic for bacteria

1. A bacterial cell consists of a *cell wall* and a compound membrane enclosing protoplasm.
2. The cell wall of bacteria is composed of a *peptidoglycan.* It is composed of *acetyle-glucosamine* and *cetyle-muramic acid.*
3. The rigid peptidoglycan layer is located between the cytoplasmic membrane and an outer multiple tract layer.
4. The latter layer is composed of *lipoprotein.* It is common in gram negative bacteria.
5. Inside the protoplasm, nucleus, vacuoles, mesosomes, polysaccharides, lipids, mitochondric granules and spores are found.
6. Many bacteria possess other intracellular membrane systems such as *mesosomes* and *chondrioids*.
7. *Mesosomes* serve for compartmentalization and integration of biochemical system.
8. Bacteria do not have characteristic *nucleus*.
9. Bacteria contain *70S ribosomes* depending upon the bacteria. Ribosomes are the sites of protein synthesis.
10. Externally bacterial cells may show *capsules, pili, fimbrae* and *flagella.*
11. Some of the bacteria possess *flagella* which are useful for *motility* of the bacteria.

12. The bacteria belonging to eubacteriales have *peritrichous flagellus* at all the sides.
13. Some bacteria transform themselves into small ovals or spheres which are highly resistant to adverse conditions. They are called as spores.
14. *Fibriae* are hair like structures that are observed as surface appendages on some bacteria. Fimbriae are common in plant pathogenic bacteria.
15. *Pili* are also hair like structures found in some bacteria. They serve as *adsorption organs* for bacteriophages. Pili mediate conjugation of bacteria.
16. Most of the plant pathogenic bacteria are rod shaped except *Streptomyces* which are mostly filmentous.
17. Majority of plant pathogenic bacteria are aerobic except; *Erwinia.*
18. Plant pathogenic bacteria required optimum growth temperature ranging between 27-30^{0}C.
19. Most of the plant pathogenic bacteria reproduce by asexual process known as binary fission or fission.
20. Most of the plan pathogenic bacteria are 'gram negative' except *Streptomyces, Clavibactor, Curtobacterium, Nocardia* and *Rhodococcus* which are 'gram positive' in Gram's reaction.

Examples of disease caused by bacteria

i) Wilts of tomato, banana, brinjal, potato
ii) Wild fire of tobacco
iii) Halo blight of beans
iv) Black rot of cabbage
v) Bacterial blight of rice
vi) Citrus canker
vii) Black arm of cotton

Fastidious Vascular Bacteria (Rickettesia Like Organisms)

Main characteristics

1. The fastidious vascular bacteria were formally known as *rickettesia like organisms* (RLOs).
2. They are usually rod shaped, aflagellate, bounded by a cell membrane and a cell wall measuring 1.0-4.0 x 0.2-0.5 μm.
3. They are confined to *phloem or xylem of the host* plant but never to both.

4. Usually transmitted by *leaf hoppers* exception: Citrus greening bacteria (*Liberobacter asciaticum* is transmitted by citrus psylla (*Diaphorina citri*).
5. Almost all fastidious bacteria known so far are *gram negative*. Exception Sugarcane ratoon stunting (*Clavibacter xyli pv.xyli*) is gram positive in nature.
6. There are two groups: one in which the fastidious bacteria occur only in phloem (phloem limited) and the other in which they occur in xylem (i.e xylem limited).
7. Xylem limited fastidious bacteria known so far have been successfully grown on nutrient media while phloem limited fastidious bacteria have not been culture so far.
8. They are sensitive to antibiotics such as *tetracycline* and *penicillin* and to high temperature.

Disease Caused by RLO

i) Plecei's disease of grapes
ii) Almond leaf scorch
iii) Alfalfa dwarf
iv) Phony peach
v) Raton stunting
vi) Plum leaf scald

Phytoplasmas

Definition

Phytoplasmas are unicellular, ultramicroscopic, walless, prokaryotic, self-replicating, highly pleomorphic, filterable organisms or entities.

Main characteristic

1. Phytoplasma *lack* rigid cell wall.
2. They are surrounded by a unit membrane, and consist of cytoplasma, ribosomes and nuclear material.
3. *Pleomorphic* or filamentous shape.
4. Size ranging from 200 to 800 nm and is less than 1 µm in diameter.
5. DNA is present in cytoplasm.
6. Genome is very small (689-1600 kb).

7. Strictly host dependent and can survive and multiply only in the sap of phloem sieve tubes or insect haemolymph; also in their eggs.
8. Phytopalsma can only be acquire and transmitted by phloem-feeding insect vectors like, *Leaf hoppers, Plant hoppers* and *Psllids*.
9. They *cannot* be grown on artificial media
10. Reproduce by *binary fission*.
11. They are sensitive to *tetracycline* but resistant to *penicillin*
12. MLOs have no flagella, produce no spore and are *gram-ve*.

Disease Caused by Phytoplasmas

i) Pear decline

ii) Grape yellows

iii) Aster yellows of vegetables

iv) Apple proliferation

v) Little leaf of brinjal

vi) Coconut lethal yellowing

vii) Elms yellows

viii) Pigeonpea witche's broom

ix) Rice yellow dwarf

x) Stolbur

xi) X.disease

xii) Citrus greening

xiii) Seasamum phyllody

xiv) Grassy shoot of sugarcane

xv) Little leaf of brinjal

Spiroplasma

Definition

Spiroplasma is a wall less mollicutes bounded by a triple layered unit membrane.

Main characteristic

1. Spiroplasma *lack* true cell wall, bounded by a single triple layered unit membrane.

2. Produce typical *'fried egg'* like colonies on agar medium.
3. Multiply by *fission.*
4. Produce helical (100-240 nm in dia., 2-4μm inlength) form in liquid culture medium.
5. Can be *cultured* on nutrient media.
6. Resistant to *penicillin* but inhibited by *tetracycline.*
7. Transmitted only through phloem – feeding insect vectors mostly through *leaf hoppers.*
8. Mostly found in the gut or haemolymph of insects, or in the phloem of infected plants.

Disease caused by Spiroplasma

i) Corn stunt

ii) White bud of maize

iii) Citrus stubborn

Viroids

Defnition

Viroid is circular, encapsulated, low molecular weight ($1.1\text{-}1.3x10^5$), self replicating, highly infectious ssRNA molecules.

Main characteristic

1. Viroid exists in *in vivo* as unencapsidated *RNA.*
2. They never contain any *protein coat* (capsid)
3. Genome of the viriod naked, *single stranded RNA* with 250-400 nucleotide, either linear or mostly circular.
4. Viroids are smaller in size than viruses i.e 50 nm or 1.1 to 1.3 X 103 molecular weight.
5. Despite its small size, the infectious RNA is replicated autonomously in susceptible cells; that is, no helper virus is required for multiplication.
6. Viroids are notable to synthesis protein and replicase enzyme required for replication.
7. Viroids replicate by direct RNA, copying in which all components required for viroid multiplication including RNA polymerase are provided by the host.
8. They cause diseases only in plants.

9. Viroids concentration and translocation is higher in growing parts of the plants (up to 0.2 mm from the apex).

Examples of disease caused by viroid

i) Potato spindle tuber

ii) Coconut cadang-cadang

iii) Citrus exocortis

iv) Chrysanthemum stunt

v) Tomato bunchy top

vi) Hot stunt viroid

vii) Avocardo sunblotch

viii) Peach latent mosic

Algae

Main characteristic

1. Algae are the *eukaryotic thallophytes* having chlorophyll as their primary photosynthetic pigments.
2. They are ubiquitous organisms abundantly present in aquatic environment.
3. They have wide range of shapes and size and the common shapes are rhizoidal, filamentous and mucilaginous.
4. The Chlorophycean (green algae) cause disease in plants.
5. Most of the algae are photoautotrophic, may reproduce either *sexually or asexually*.
6. Algae reproduce *asexually* by *binary fission*, fragmentation and formation of spores and *sexually* by formation of *zygotes*.
7. They spread through an air borne sporangia and produces 300 spores.
8. They entry in to the host through stomata and other natural openings.

Example of disease caused by algae

- Red rust of tea, coffee, mango citrus, and guava (*Cephaleurous parasitica, C.minimum, C.coffeae).*

Protozoa

Main characteristic

1. Protozoa are single celled, non photosynthetic, eukaryotic animal like organisms.
2. They are widely distributed in nature particularly in aquatic, environment or moist habitats.
3. Their size ranges from 2-3 micron.
4. They *lack rigid cell wall* and *do not contain chlorophyll.*
5. They make their movements with the help of *cilia, flagella or pseudopodia.*
6. Most of the protozoa are free living but some are symbiotic or parasitic causing diseases in plants, animals and humans.
7. Free living protozoa are commonly found in fresh water, salt water, sand, soil and decaying organic matter.
8. Protozoa reproduce *sexually or asexually.*
9. Asexual reproduction is either by *binary fission* or by budding or by both.
10. Sexual reproduction is by fusion of two gametes.
11. They have characteristics ability to regenerate the lost parts of the body. For example, Genera: Amoeba, Paramecium etc.

Example

1. Phloem necrosis of coffee (*Phytomonas leptovasorum*)
2. Hart rot of coconut and oil palm (Phytomonas)
3. Sudden wilt of oil palm
4. Empty root of Cassava

Prions

Main characteristic

1. *Proteinaceous* infectious particles that *lack nucleic acid.*
2. It is the smallest known protein often considered to be the cause various infectious diseases of the nervous system.
3. Prions are some times called slow viruses because of their slow effect.
4. Prions are transmissible particles .
5. No plant diseases are reported to be caused by prions.

Diseases caused by Prions

- Mad cow disease & Creutzfeldt- Jakob disease in humans
- Scrapie disease in sheep.

Phanerogamic Parasites

The parasitic flowering plants are also important pathogens. There are few seed plants, which are parasitic on living plants and are called parasitic higher plants or **Phanerogamic parasites.** These parasitic higher plants attack some valuable crops and trees causing considerable losses. More than 2500 species of higher plants are known to live parasitically on other plants.

Classification

i) Complete Parasite (Holoparasites)

a) Root : Rafflesia, Orobanchea (Broom rape)

b) Stem: Cuscuta (Dodder Amarvel)

ii) Partial Parasites (Semi Parasites)

a) Root: Striga (Sandle wood witch weed)

b) Stem : Loranthus (Banda or Deudrophae)

Nematode

Definition

Nematodes may be defined as *bilaterally symmetrical, triploblastic, pseudocoelomic unsegmented invertebrates with four hypodermal chords, a tri-radiate oesophagus,circum oesophageal nerve ring and a definite tail but lack circulatory muscles and specialized organs for locomotion and respiration.*

Main characteristic of plant parasitic nematodes

1. Plant parasitic nematodes are elongated, cylindrical, *bilaterally symmetrical*, unciliated worms with body tapering towards both the ends.
2. The body is covered by a transparent, tough, *resistant cuticle* that is secreted by a layer of hypodermal (epidermal) cells lying just beneath it.
3. Plant parasitic nematodes have a protrusible hypo-pharyngeal needle like knobbed *stylet (spear)* at the anterior end of the body, which is used to *suck* the cell cytoplasm from the host plants.

4. Nematodes possess a false body cavity or *pseudocoelom* i.e a cavity present between mesoderm and endoderm. Pseudocoelom is fluid filled and most part of it is occupied with intestine and components of reproductive system.
5. Oral aperture (mouth) is terminal and is surrounded by lips which contain sensory *cephalic papillae*. All other body openings like excretory pore, vulva and anus (in females) and cloacal aperture (in males) are ventrally located.
6. Musculature consists of smooth longitudinal muscle cells located between the hypodermal chords and are directly connected to longitudinal nerves. Circulatory muscles are absent.
7. *Digestive, reproductive, nervous* and *excretory systems* are *present*. Well defined *respiratory* and *circulatory* organs are *absent.*
8. They are usually *dioecious* with tubular reproductive organs. One or two gonads may be present.
9. They do not have any specialized structures for locomotion.
10. The life cycle of a plant parasitic nematode has six stages; egg, four juvenile stages and adult.
11. Nematodes belong to the animal kingdom, and phylum *Nematoda.*
12. Most of the important parasitic genera belong to the order *Tylenchida* and few under Dorylaimida.

Examples of highly damaging plant parasitic nematodes

1.	Root knot	*Meloidogyne* spp
2.	Cyst	*Heterodera* and *Globodera* spp
3.	Root lesion	*Pratylenchus* spp.
4.	Spiral	*Helicotylenchus* spp.
5.	Burrowing	*Radopholus similis*
6.	Bulb and stem	*Ditylenchus dipsaci*
7.	Reniform	*Rotylenchulus reniformis*
8.	Dagger	*Xiphinema* spp
9.	Bud and leaf	*Aphelenchoides* spp

Questions

1. Which green algae are plant pathogens ?
2. Which are the common plant pathogenic species of Cephaleuros?
3. Write the classification of Cephaleuros?
4. What is the sign and symptoms of red rust of tea?

5. What is the name of lichenized state of Cephaleuros?
6. Write three groups of Fastidious vascular colonizing bacteria?
7. Describes the unique features of the classification of the wall less mollicutes?
8. Write short notes on any four:

I. Spiroplasma

II. Clavibacterium

III. *Candidatus liberator*

IV. Phytoplasma

9. Why is the designation RLO now abondaned?
10. Why are aphids unable to transmit mollicutes?
11. Why are Phytoplasma and Liberatobacter treated as Candidatus genera?
12. Why are Phytoplasmas considered as basically insect parasites?

9

Diseases and Symptoms Due to Abiotic Causes

Definition

Plant disease in which no pathogens or parasitic is associated with the cause is known as Non-parasitic disease. They are also called as *non-infectious or physiological disorders.*

General Characteristics

1. Occurs in the absence of pathogen and therefore, *cannot be cultured* or *transmitted.*
2. Caused by the *lack or excess* of something that supports life from diseased to healthy plants.
3. Effect plant in *all stages of their lives*, such as seeds, seedlings, mature plants or fruits.
4. Symptoms may range from *slight to severe* & affected plants may even die.
5. These disease cause damage in field, in storage or at the market.

Factors responsible for physiological disorder

I Temperature
II Moisture
III Oxygen
IV Atmospheric impurities
V Soil toxicity
VI Herbicidal injury
VII Nutritional deficiency

I. Temperature

i) Effects of low temperature

1. Causes greater damage to crops than high temperature.
2. Frost injuries involved killing of buds of peach, cherry, killing of flowers, young fruits and sometimes succulent twinges of most trees.
3. Low winter temperature may kill young roots of trees such as apple and may also cause canker development and bark splitting.

ii) Effects of high temperature

1. Plants are normally injured quickly.
2. Responsible for sun scald injuries appearing the sun exposed sides of fleshy fruits and vegetables. For example. Sun scald of apple, tomato, onion- bulbs, peppers, potato tubers, canker of linseeds.

II. Effects of moisture

i) Effects of low moisture

1. Plant generally remain stunted.
2. Pale green to light yellow leaves,
3. Plant dry and wilt.

ii) Effects of high moisture

1. Decay or rotting of fibrous root of plants resulting into wilting.
2. Accumulation of toxic materials around the root and base of stem.
3. Non availability of nutrients.

For example. Tip burn of paddy.

III. Inadequate Oxygen

1. Disintegration of cells due to enzymic action.

For example. Black heart of potato.

IV. Unfavourable Light

1. Retards chlorophyll formation.
2. Promotes slender growths with long internodes
3. Pale green leaves, spirally growth and premature drop of leaves and flower. This condition is known as "*Etiloation*".

V. Atmospheric Impurities

1. Presence of injurious gases in the atmosphere may cause definite injury to plant and plant parts.
2. More severe and wide spread damage is caused to plant in the fields by chemicals such as hydrogen fluoride, nitrogen dioxide, ozone, sulphur dioxide, peroxyacetyl nitrates. For example. Black tip of mango .
3. Fruit on trees in close proximity to brick kilns may bear necrotic lesions and become useless for sale and consumption.
4. The smoke of kills polluted the air with toxic gases like sulphur dioxide which caused necrosis of tissues.

VI. Soil mineral toxicity

1. Excessive amount of sodium salt especially sodium sulphate, sodium chloride, and sodium carbonate raise the PH of the soil and cause alkali injury i.e chlorosis, stunting etc.
2. Boron, copper and manganese often implicated in mineral toxicity disease.
3. Excess manganese is known to cause crinkle leaf disease in cotton.
4. Excess boron is toxic to many vegetable and trees.

VII. Herbicidal injury

1. Several of the most common plant disorder seems to be the result of extensive use of herbicides.
2. Increasing number of herbicides in use for general or specific weed control has created problems.

VIII. Nutritional deficiencies or Disorder in Plants

1. Deficiency of minerals *viz*, *Nitrogen, Phosphorus, potash, manganese , zinc, copper, iron, magnesium, boron,* etc. results in disorders in plant metabolism and cause *hunger signs* in the crops
2. Excess and deficiency of minerals disturbs nutritional balance needed for good metabolism in the plant, hence hindering the effect of essential element.
3. The deficiencies and excess of minerals also reduce the resistance of plant to fungal, bacterial and other diseases.
4. The kind of symptoms produced by deficiency of a certain nutrient depend primarily on the functions of that particular element in the plant.
5. Nutrient mobility plays an important role in determining the site of the deficiency symptoms.
6. For *highly mobile* elements like *nitrogen and potassium*, the deficiency symptoms appear predominantly in *older* and mature leaves.
7. Deficiency of *poorly mobile* elements, like *calcium, boron* and *iron* show up in *younger* leaves.
8. The symptoms of deficiency of *moderately mobile* elements like *magnesium* and *sulphur,* are uniformly spread all over the plant.

IX. Symptoms of mineral deficiency

The symptoms of mineral deficiencies is recorded in the following table:

S.No.	Deficient element	Symptoms
1.	Nitrogen	Poor growth, chlorosis of leaves, delayed flowering and fruiting
2.	Phosphorus	Poor growth, blue/green, but not yellow, colour of leaves, oldest leaves affected first
3.	Potassium	Brown scorching and curling of leaf tip, yellowing of leaf veins
4.	Calcium	Curling and tip burn of leaves, stunted growth, abnormal development of meristmatic tissues, and eventual death of buds and root tips. Also causes ***bitter pit in apple***.
5.	Magnesium	Interveinal chlorosis, giving a mottled appearance to leaves,similar to virus infection
6.	Sulfur	Chlorotic leaves, petioles and veins distinctly red.
7.	Iron	Chlorosis of leaves, young leaves appear bleached, symptoms similar to manganese deficiency
8.	Manganese	Chlorotic leaves with green veins
9.	Zinc	Younger leaves yellow, pitting on lower leaf surface, leaves and internodes shortened, giving a rosette-like look to plants; guttation increased
10.	Copper	Leaves pale and curled; petioles droop down
11.	Boron	Drying of growing tip, bushy stunted growth, & internal tissue breakdown.
12.	Molybdenum	Chlorosis of leaves; symptoms similar to nitrogen deficiency. Also causes whiptail disease in cauliflower.

Table : Plant disease due to lack of minerals.

S.No.	Deficient Nutrient	Disease
1.	Nitrogen (N)	Red leaf of cotton
2.	Phosphorus (P)	Dwarfening of cotton
3.	Potassium (K)	Cotton rust, leaf spot of alfa alfa
4.	Magnesium (Mg)	Sand drown disease in tobacco
5.	Boron (B)	Internal cork of apple, Cracked stem of celery, Black tip of mango, Heart rot of sugarbeet, Brown heart of cabbage and turnip, Internal brown spot of sweet potato, Terminal bud breakdown of tobacco, Fruit pitting and dieback of olive, Hollow stem of brassica
6.	Copper (Cu)	Reclamation disease of oats, Diebacks of citrus, Wither tip of apple
7.	Manganese (Mn)	Pahala blight of Sugarcane, Marsh spot of garden pea, Grey speck of oats
8.	Zinc (Zn)	White bud of corn, Little leaf of apple, Bronzing of twigs, Rosette of fruits trees, Mottle leaf of citrus, Khaira disease of rice
9.	Iron (Fe)	Green netting of citrus
10.	Molybdenum (Mo)	Whiptail disease of cauliflower
11.	Calcium(ca)	Blossom end rot of tomato, Black heart of celery, Wither tip of flax.

Questions

1. How nutrient mobility determines the site of symptom development?
2. Which elements become deficients in acidic soils?
3. Why is iron not easily available to the plants?
4. Distinguish between :
   I. Iron efficient and Iron deficient plants
   II. Freezing and Chilling injury
   III. Primary and secondary air pollutants
5. Write notes on i. Microbial inoculants, ii. Iron efficient plants iii. Plant disorder caused by Zn. iv. Disorder caused by Fe
6. Write short notes on i. Disorder caused by B ii. Disorder caused by N iii. Disorder caused by Mo iv. Disorder caused by Ca.

10

Fungi: General Characters, Definition of Fungus, Somatic Structures, Types of Fungal Thalli, Fungal Tissues, Modifications of Thallus & Reproduction

a) Definition

Fungus is Latin word meaning **'mushroom'**

Alexopoulus and Mims 1979 defined fungus as *eukaryotic, achlorophyllous organism generally produced by sexual or asexual method and whose filamentous branched somatic structure is typically covered by cell wall.*

b) General Characteristics of Fungi

1. **Thallus:** The body of the fungus is called as *Thallus*, which is without stem root and leaves. A single thread like filament is called as *hypha.* A group of hypha constituting the body of fungus is called as *mycelium* may be *septate* or *aseptate*. i.e coenocytic
2. **Cell Wall:** Cell wall is well defined, typically made up of *chitin* (*Cellulose* in *oomycetes*)
3. **Nutrition:** *Heterotrophic* i.e. Photosynthesis lacking and absorptive
4. **Nuclear Status:** *Multinucleate*, well defined structures i.e. nuclear membrane, nucleolus, and chromatin material etc.
5. **Sexuality:** *Asexual* and *Sexual*
6. **Life Cycle:** Simple to Complex
7. **Sporocarps:** Microscopic or macroscopic & showing limited differentiation.
8. **Distribution:** Cosmopolitan

c) Somatic Structures

1. Haustoria
2. Appressoria

1. **Haustoria:** A *modified mycelial branch, knob like*, makes intimate contact with plant cell and *absorbs food*.
2. **Appressoria:**

- These are *localized swellings* of the tip of *germ tube* that develop in response to contact with the host.
- These are special structures for attachment in the early stage of infection.

d) Types of Fungal Thalli

There are two types of thalli in fungi

1. Homothallic
2. Hetrothallic

1. Homothallic Fungi

- Male and female sex organs or both the gametes are produced on the *same thallus*, E.g. Powdery mildew of Mung – *Sphaerotheca fulginae.*

2. Hetrothallic Fungi

- If male and female sex organs or both the gametes are produced on the *different thallus*, E.g. Rust fungi.

e) Fungus Tissue

There are three types of tissues found in fungi

1. Prosenchyma
2. Pseudoparenchyma
3. Plectenchyma

- **Prosenchyma**: The loosely woven tissue in which the component hyphae with elongated cells lie more or less parallel to one another is called prosenchyma.
- **Pseudoparenchyma**: The fungal tissues which are closely packed, in the form of more or less isodiametric or oval cells resembling the parenchyma cells of higher plants are called pseudoparenchyma.

Both prosenchyma and pseudoparenchyma compose various type of vegetative and reproductive structures.

- **Plectenchyma**: The organized fungal tissues are called *Plectenchyma.*

f) Modification of Mycelium or Thallus

Rhizomorph

- The mycelium of some fungi forms *thick strands*, in such strands the hyphae lose their individuality and form complex tissues.
- Its function is believed to be the transportation of water across dry areas.
- They belongs to subdivision Basidiomycotina. e.g. *Arnillaria millea.*

Stroma or Stromata

- It is a *compact mass of the hyphae* and appears as pseudoparenchymatous tissue and contain fruiting body of the fungus. E.g. ergot of bajara.

Sclerotium

- Asexual, thickwalled, multicelled, hard, rounded overseasoning structure.
- It may remain dormant for long periods and germinate under favourable conditions.

Dormant Mycelium

- It is the mycelium which hibernates in the host tissue to tide over unfavourable conditions. E.g. Loose smut of wheat, Downey mildew of grape.

Gemmae

- These are the chlamydospore produced in lower fungi whose walls are thinner.
- They occur either singly or in chains, becomes separated after maturing.
- Gemmae break free from the mycelium and disperse in water.
- e.g. Mucor sp, Saprolegnia sp.

g) Reproduction

Reproduction is the *formation of a progeny by either sexual or asexual means*. Spore is an unit of reproductions.

Types of reproduction

- Asexual reproduction
- Sexual reproduction

Asexual Reproduction

This method of reproduction is characterized by *production of identical individuals without the union of the sex organs.*

Methods of Asexual Reproduction

1. By fragmentation
2. By arthrospores or oidia
3. By budding
4. By binary fission
5. By production of spores/conidia

Asexual Reproduction

1. By fragmentation

- Breaking off of parts of the mycelium through external forces.
- Such pieces of mycelium under favorable conditions can start a new individual.

2. By arthrospores or oidia

- Cells of the hyphae at the distal end round off and separate in basipetal succession.
- On germination, the arthrospores give rise to new fungus colonies.

3. By Budding

- Production of a small outgrowth from a parent cell.
- Bud increases in size while still attached to the parent cell.
- It eventually breaks off and forms a new individual .
- E.g. Rust, Smut fungi & yeast fungi.

4. Binary Fission

- Fission can occur through the simple splitting of a cell into two daughter cells by constriction.
- This is found among the bacteria generally, but some fungal yeasts may do this also.

5. Production of conidia

- Conidia produced on conidiophores in basipetal (oldest at the top and youngest at the bottom) and acropetal (oldest at the bottom and youngest at the top).

Sexual Reproduction

Sexual reproduction in fungi involves the *union of two compatible nuclei.*

Methods of Sexual Reproduction

a) Planogametic Copulation
b) Gametangial Contact
c) Gametangial Copulation
d) Spermatization
e) Somatogamy
f) Heterokaryasis
g) Dikaryotization

The process of sexual reproduction consists of three distinct phases.

1. **Plasmogamy:** It is *union of two protoplasts* brings the nuclei close together within the same cell.
2. **Karyogamy:** Actual fusion of *two haploid nuclei* brought together as a result of plasmogamy , Karyogamy immediately follows plasmogamy.
3. **Meiosis:** The *nuclear fusion is followed by meiosis*. Meiosis reduces the number of chromosomes to haploid.

Methods of Sexual Reproduction

a) Planogametic copulation (Gametogamy)

This involves the *fusion of two naked gametes one or both of which are motile.* Motile gametes are called planogametes. Depending on the size and motility of the fusing gametes, there are three types

- **Isogamy** : *Union of gametes that are similar in shape and size.* e.g. *Synchytricum, Olpidium.*
- **Anisogamy** : *Union of gametes that are morphologically similar but differ in size.* e.g. Genus-Allomyces.
- **Heterogamy**: *Union between a motile male gamete with a non motile female gamete* is known as heterogamy.

b) Gametangial Contact

In this method, *two gametangia of opposite sex come in contact, and one or more gamete nuclei migrate from the male to the female through especially developed fertilization tube serves as a passage for the male nuclei.* The zygote formed is called Oospore. e.g. Oomycetes, Ascomycetes.

c) Gametangial Copulation

Entire protoplast is transferred into the female gametangia as involes the fusion of two protoplast in a common cell. Sex gametes are indistinguishable or morphologically identical. Copualtion occurs either by complete fusion of two protoplast. It is common in class Trichomycetes and Zygomycetes.

d) Spermatization

It involves the formation of small spores or seed like structures or *spermatia. Spermatia acts as male gamete which is uninucleate or spore like and carried out by wind, insects to the retentive hypha* (Female gametangium). A pore developed at the place of contact and the contents of the spermatia passes into receptive hypha which serves as a female organ. e.g. Pycniospores and Receptive hypha in rusts.

e) Somatogamy

No gametes are involved. *Vegetative hypha itself acts as a male and female gamete and bring about sexual reproduction.* e.g. Smut fungi.

f) Dikaryotization

It is accomplished through *migration of nuclei from one cell to another cell of vegetative hypha, often through mechanism of clamps.* The two nuclei remain in pair and divide as such and only fuse prior to the formation of spores. No special sex cells are produced. Clamp connections are formed during nuclear division. E.g. Class-Basidiomycetes.

g) Heterokaryosis

The phenomenon of the *existence of genetically different kinds of nuclei in the same individual is called heterokaryosis,* and the individuals that exhibit it are heterokaryotic. Heterokaryons may originate in a fungus thallus in four ways:

- By the introduction of genetically different nuclei into a homokaryon, a somatic cell in which all nuclei are similar.
- By the germination of a heterokaryotic spore, which will give rise to heterokaryotic soma.
- By mutation in a multinucleate, homokaryotic structure and the subsequent survival, multiplication, and spread of mutant nuclei among the wild type nuclei, and
- By fusion of some nuclei in a haploid homokaryon, and the subsequent survival, multiplication, and spread of the diploid nuclei among the haploid.

Special type of Sexual Reproduction

Parasexuality: Some fungi do not go through a true sexual cycle as described. They may derive the benefits of sexual recombination through a process known as parasexuality.In this process, *plasmogamy, karyogamy and haplodization take place, but not a specified points in the thallus or the life cycle.* The parasxual cycle involves the following steps:

1. Formation of heterokaryotic mycelium.
2. Nuclear fusion and multiplication of the diploid nuclei.
3. Mitotic crossing over during the division of the diploid cells.
4. Sorting out of the diploid strains.
5. Haplodization.

Questions

1. What is fungus? Write down their general characters and method of reproduction.
2. Discuss in details about sexual and asexual reproduction in fungi.
3. What is fungi? Write down the characters of fungi.
4. Write short notes on i.Parasexuality ii.Heterokaryosis iii. Dikaryotization iv. Fungal tissue.
5. Differentiate between:

a) Haustoria and Appresoria.

b) Homothallic fungi and Heterothallic fungi.

c) Gametangial contact and Gametangial copulation.

d) Cytokinesis and Karyokinesis.

6. Differentiate between.

a) Sexual reproduction and Parasexual reproduction.

b) Isogamy and Anisogamy.

c) Spermatization and Somatogamy.

d) Rhizomorph and Sclerotium.

11

Symptoms of Plant Diseases

Introduction

Plant pathogens induce diverse reactions in the body of their hosts. This result in creation of abnormalities which become visible on the plants. A visible or otherwise detectable expression of abnormal physiology, development, or behaviour in a plant resulting from disease is called symptoms. Normally, a distinction is made between symptoms and signs *A symptom of disease is expressed as a reaction of the host to a causal agent, whereas, a sign is evidence of disease other than that expressed by the host. Signs are usually the structures of the pathogen.* A disease is first noticed by the presence of symptoms and /or signs, and recognition of specific type of symptoms or sign aid in the eventual diagnosis of the disease.

Classification of Symptoms

Symptoms are classified into three general categories:

A. **Necrotic symptoms:** Those symptoms that results from termination of function leading to death.

B. **Hypertrophy/Hyperplasia symptoms:** Overdevelopment; *Hyperplasia*; An abnormal increase in number of cells in a particular tissue or organ. *Hypertrophy*; An abnormal increase in the size of a plant or plant part generally due to an abnormal increase in the size of the cells.

C. **Hypoplastic:** Under development; An abnormally small number of cells in a particular tissue or organ.

A. Necrotic symptoms

1. **Anthracnose:** It is a sunken ulcer like lesion of necrotic cells on the infected part of the host plant.
2. **Blasting:** The failure to develop fruit.
3. **Blight:** The sudden drying and browning of whole leaves, shoots or branches.
4. **Blotch**: Large, irregular lesions on leaves, shoots and stems.
5. **Canker**: Necrotic, often sunken lesions in the bark or cortex of the stem and roots of especially woody plants.

6. **Damping-off:** A symptom complex characterized by rapid dying, browning, and rotting of germinating seedlings. Shoots may be killed before emergence, stems may be attacked in the root collar region causing shoots to fall over, roots may be destroyed, or cotyledons may be attacked.
7. **Decay:** Disintegration of dead tissues.
8. **Dieback:** The progressive drying, shrivelling, and browning of twigs or branches from the tips inward toward the trunk.
9. **Hydrosis:** A water soaked, translucent appearance of leaves, fruits and green stems due to the extrusion of water from the cells into the intercellular spaces.
10. **Scald:** Blanching of the epidermis and adjacent tissues.
11. **Scorch :**A sudden drying and browning of large, indefinite areas on leaves and fruits. Also damage to bark resulting in drying and death.
12. **Shot hole**: Circular hole, in leaves resulting from the dropping out of the central necrotic areas of spot.
13. **Spot:** Circular, areolate, or irregular discolored and dead areas on leaves, fruits, or green stems.
14. **Streaks or stripes**: They are prominent symptoms consisting of an elongated but relatively narrow lesions. These streaks or stripes are usually some shade of brown colour.
15. **Ring Spots**: Appearance of sometimes chlorotic but mostly necrotic rings on the leaves or also on the fruits and stem.Ring spots are common type symptoms produced by viral pathogens
16. **Rot :** The disintegration and decomposition of dead tissues.
17. **Wilting:** A flaccid appearance of leaves and shoots resulting from a temporary or permanent loss of turgor due to excess transpiration by the leaves and shoots.
18. **Yellowing:** The loss of green color from chlorophyllous tissues, due to the destruction of the chlorophyll and/or degeneration of the chloroplasts, which unmasks yellow pigments.

B. Hypertrophy/Hyperplasia (Overdevelopment) symptoms

1. **Anthocyanescence:** A reddish or purplish coloration of leaves resulting from abnormal development of anthocyanin pigments.
2. **Callus:** Tissue overgrowth produced in response to injury or other irritation and which tends to cover a wound, canker, etc.

3. **Enations:** Enations are outgrowths generally occurring in veins or midrib on the lower surface of leaves. They may be small, large, papillate, or spin like in shape and also vary in number.
4. **Leaf curl**: Abnormal bending or curling of leaves or shoots due to localized overgrowth on one side or in certain tissues.
5. **Fasciation :** Flattening or cohering of organs such as stems, flowers, and roots. The cause is unknown, but in some cases the condition can be propagated.
6. **Witches' broom:** A type of overgrowth in which there is an abnormal bush like development of many weak shoots or roots.
7. **Sarcody:** Abnormal swelling of tissues above girdled branches or stems.
8. **Scab:** A limited, more or less circular, raised, and sometimes roughened lesion on fruits, tubers, leaves, and stems.
9. **Tumor:** Local swelling on any part of the plant, usually woody roots, stem, or branches, usually resulting from stimulation of the plant meristem by the pathogen.
10. **Virescence:** The process in which a normally white or colored tissue develops chlorophyll and becomes green.
11. **Warts:** Warts are protuberance developed on tubers and stems generally giving warty or discolored cauliflower like appearances.

C. Hypoplastic (Under development) symptoms

1. **Chlorosis:** Yellowing of green tissue due to chlorophyll destruction or failure of chlorophyll formation.
2. **Dwarfing:** Subnormal size of a plant or some of its organs.
3. **Etiolation:** A symptom complex in which the major symptoms are dwarfing of foliage and inflorescence, spindly stem growth, and chlorosis.
4. **Mosaic:** Pale green mottling of leaves.
5. **Russeting:** Rough or corky surfaces formed where they do not normally occur.
6. **Suppression:** The complete prevention of organ development.
7. **Leaf narrowing:** The infected leaves generally become narrow due to reduced growth of laminar tissue but the veins and midrib remain normal in growth.
8. **Leaf curling:** These symptoms represent irregular and extensive wrinkling and furrowing of leaves due to reduced growth of vein in comparison to the

growth of laminar tissue resulting in sunken veins and raised laminar tissue leading to the curling of the leaves.

Classification of Signs

Signs are divided into three general categories:

A. Vegetative structures

B. Reproductive structures

C. Disease products

A. Vegetative structures

Mycelium, haustorium, pathogen cells, rhizomorph, sclerotia, etc are structures generally observed.

B. Reproductive structures

Apotheica, Acervuli, asci, basidium, cleistothecium, conidiophores, mildews, perithecia, pycnidia, sporangium, spores, sporodochium etc are the structures commonly observed.

C. Disease products

In various bacterial diseases *viz,* bacterial blight of rice, masses of *bacterial ooze* out to the surface of the affected organs where they may be seen as drops of various sizes or as thin smeer over the surface. In some cases some what raised, black coated fungus bodies appearing as a flattened out drop of tar on leaves appear.

Questions

1. Clarify 'disease syndrome'.
2. What do you understand by 'masked symptoms'?
3. What are the shot-holes in necrotic symptoms?
4. Why the symptoms manifested on potato leaves due to *Phytophthora infestans* are called 'blight'?
5. Differentiate between 'die-back' and 'damping-off with examples.
6. Differentiate between 'downy mildew' and 'powdery mildew' symtoms.
7. Name three diseases (with causal agents) in which the symptoms appear on hosts in the form of gall/tumour.
8. Clarify the terms 'vein-chlorosis, 'vein-clearing', and 'vein-banding'.

9. What is necrosis? Describe any five types of necrotic symptoms with the help of suitable illustrations.
10. Attempt the following:
a) Rusts and smuts
b) Hypertrophy and hypotrophy
c) Chlorosis
d) Rots.
11. Define disease symptoms and sign.Describe in brief the different types of symptoms produced in plants giving examples.
12. What are the disease signs produced in plants due to vegetative and reproductive structures of the visible pathogen.
13. Describes in detail the difeerent types of hyperplastic symptoms of diseases produced by plant pathogens.

12

Nomenclature, Binomial System of Nomenclature, Rules of Nomenclature Classification of Fungi Key to Divisions, Sub-divisions, Orders and Classes

Taxonomy

Taxonomy is the science that deals with the *identification, nomenclature* and *classification* of organisms.

Systematics

It is *scientific study of organisms* with the ultimate object of *characterizing* and *arranging* them in an orderly manner.

Classification

Classification is the *ordering* of organisms into groups and systems, based on their *morphological, phylogenetic* and *physiological* relationships.

Nomenclature

It is the system of assigning *names* to the *taxonomic groups or organism* according to international rules.

Binomial System of Nomenclature

Binomial system of nomenclature was developed by *Carlous Von Linnaeus* which is now universally used. As per the binomial system the *name of organism* is composed of *two words*. The *first word* designates the *genus* and the genus name is always *capitalized.* The *second word* designates the *species* and its name is *not capitalized.* Binomials when written are *underlined* and when printed *italicized*. E.g. *Erwinia coli*, in which *Erwinia* is the *genus* and *coli* is the *species*.

Rules of Nomenclature

Following rules should be observed while righting of binomial.

1. The name of the genus should always be *capitalized.*
2. Species name should *not be capitalized.*
3. Binomial when written should always be *underlined separately*; when printed *italicized.*
4. The name or abbreviated name of the scientist describing the species for first time should be written after binomial. E. g. *Pseudomonas syringae* Val Hall.
5. If the name is revised, the name of the original describer should be written in *bracket* followed by the name of the revising scientist. E.g. *Xanthomonas compestris pv oryzae* (Dye).
6. To avoid confusion the same binomial should not be used to name two different species.
7. The year in which organism was described should be written after the name of the author or scientist.

Classification of fungi

The following suffixes attached to the scientific name of a taxon indicates the various ranks in the system of classification based on the one proposed by Ainsworth (1973).

Taxa	Standard endings	Description
Super kingdom	Eukaryota	-
Kingdom	Now Fungi	A group of similar divisions
Sub-kingdom	Mycota	-
Division	mycota (suffix)	A group of similar classes
Sub-division	mycotina (suffix)	-
Class	mycetes(suffix)	A group of similar orders
Sub-class	mycetidae (suffix)	-
Order	ales (suffix)	A group of similar families
Family	aceae	A group or collection of similar genera
Genus	-	A group or collection of similar or closely related species.
Species	-	Collection of strains having similar characteristics
Strain	-	It is population of organism that descends from a single organism or pure culture isolate.

No standard suffixes are given for species, genus, subspecies, and varieties. In the classification given below only those groups have been emphasis which contain plant pathogenic fungi or which are important from taxonomic point of view.

Key to divisions, sub-divisions, classes and orders

Kingdom : Protista (Eukaryotic)

Sub-Kingdom : Mycota (fungi)

Key to Divisions of Mycota

1. Division : Myxomycota

- Plasmodium or pseudoplasmodium are present

2. Division: Eumycota

- Plasmodium or pseudoplasmodium are absent. Assimilative phase typically filamentous.

Division : MYXOMYCOTA

Sub-division : Myxomycotina

Class : Acrasiomycetes : Assimilative phase free living amoeba which unite as pseudoplasmodium before reproduction

Class : Hydromyxomycetes : Assimilative phase of a plasmodium, net plasmodium

Class: Myxomycetes : Assimilative phase plasmodium not a network, Saprobic or free living

Class: Plasmodiophoromycetes : Assimilative phase plasmodium not a network, endoparasites on plant.

Division: EUMYCOTA

Key to sub divisions

- **Mastigomycotina:** Sexual spore *Oospores*
- **Zygomycotina:** Perfect (sexual) state present as *Zygospores*
- **Ascomycotina:** Perfect (sexual) state present as *Ascospores*
- **Basidimycotina:** Perfect (sexual) state present as *Basidiospores*
- **Deuteromycotina:** Perfect (sexual) state *absent*

Subdivision: Mastigomycotina

General characters

- Majority of them are with filamentous hyaline *coenocytic mycelium.*
- Asexual spores called *zoospores.*
- Sexual spore called *Oospore.*

Classification

The Mastigomycotina consists of four classes

- Chytridiomycetes (with *posteriorly uniflagellate zoospores*)
- Hypochytridiomycetes (with *anteriorly uniflagellate zoospores*)
- Plasmodiophoromycetes (with *anteriorly biflagellate zoospores*)
- Oomycetes (*biflagellate zoospores*)

Class: Chytridiomycetes

- Characterized by the single posterior whiplash flagellum of their zoospores. It is divided into three orders.

Order: Chytridiales

Family: Olpidiaceae

- True mycelium absent

Genus: *Olpidium*

Class: Plasmodiophoromycetes

Order: Plasmodiophoromycetales

Family: Plasmodiophoraceae

- All are obligate parasites of higher plants, algae and fungi.
- It form a wall less, naked *plasmodium* as the somatic phase.

Genus: *Plasmodiophora*

Class: Oomycetes.

Order: Peronosporales

Family: Pythiaceae

- Sporangiophores similar to the vegetative hyphae or if different then of indeterminate growth.

Genus: *Pythium* and P*hytophthora*

Family: Albuginaceae

- Sporangiophores strikingly different from vegetative hyphae, slender or thick, variously club-shaped, arranged in a layer, and bear sporangia in chain at the tip.
- These are obligate parasites.

Genus: *Albugo.*

Family: Peronosporaceae

- Sporangiophores strikingly different from vegetative hyphae, Sporangia, singly or in clusters borne at the tip of characteristically branched sporangiophores of determinate growth.
- These are obligate parasites.

Genus: Sclerospora Pernospora Plasmopara

Order: Saprolegniales

Family: Saproleginaceae

- A peculiar character of this group is *diplanetism.*
- **Genus**: *Saprolegina* and *Achlya*

Subdivision -Zygomycotina

General characters

- Cell wall is mainly composed of *chitin.*
- Sexaul spore called *zygospore.*

Class-Zygomycetes

i) **Mucorales** : Asexual reproduction by *spores* or occasionally by *conidia.*

ii) **Entomophthorales**: Chiefly parasitic on insects.

iii) **Zoopagales**: Chiefly parasitic on insects.

Only Mucorales are discussed here.

Order: Mucorales

Family: Mucoraceae

Genus: Mucor and Rhizopus

Subdivision: Ascomycotina

General characters

- Sexual spores are the *ascospores.*, usually eight in numbers.

- Cell wall is made up of chitin.
- The fruiting body ascocarps are either *cup or saucer shaped* (*apothecium*), *flask shaped (perithecium)*, or *closed (cleistothecium).*

Key to the classes of Ascomycotina

- Ascocarps and ascogenous hyphae absent, thallus yeast-like-**Hemiascomycetes**
- Ascocarps and ascogenous hyphae present, thallus mycelial: asci bitunicate, ascocarp an ascostroma-**Loculoascomycetes**
- Asci typically unitunicate, if bitunicate, ascocarp as apothecium: ascocarp as cleistothecium, asci evanescent and scattered-**Plectomycetes**
- Asci regularly arranged as basal or peripheral layer in the ascocarp Insect parasites-**Laboulbeniomycetes**
- Ascocarp perithecium, Not insect parasites-**Pyrenomycetes**
- Ascocarp apothecium-**Discomycetes**

Class: Hemiascomycetes

- It is divided into three orders:

Order : Endomycetales: Asci developing parthenogenetically from a single cell or directly from a zygote formed by population of 2 cells.

Family : Saccharomycetaceae

Genus : *Sacchromyces, Schizosaccharomyces*

Order : Taphrinales: Asci arise from binucleate ascogenous cells formed by breaking of cells from hyphae.

Family : Taphrinaceae

Genus : *Taphrina*

Order : Protomycetales: Asci developing in a compound spore sac (syn ascus), produced singly from thick walled chlamydospores.

Family: Protomycetaceae

Genus: *Protomyces*

Class: Loculoascomycetes

It comprises the following 5 orders

Myriangiales, Dothideales, Pleosporales (Pseudosphaeriales), Hemisphaeriales (Microthyriales) and Hysteriales.

Order: Myriangiales

Family: Myriangiaceae

Genera: *Elsinoe, Myriangium*

Order: Dothideales

Family: Capnodiaceae

Genera: *Capnodium, Limacinia*

Family: Dothideaceae

Genera: *Mycosphaerella, Guignardia*

Order: Pleosporales

Family: Venturiaceae

Genera: *Venturia*

Class: Pyrenomycetes

Order: Erysiphales

Family: Erysiphaceae

It has the following genera.

1. Ascocarps present

A. Mycelium superficial

1. Ascocarp containing one ascus only

a) Perithecial appendages simple, myceloid -*Sphaerotheca*

b) Perithecial appendages dichotomously branched -*Podosphaera*

2. Ascocarp containing many asci

a) Perithecial appendages simples, myceloid -*Erysiphe*

b) Perithecial appendages coiled at the top –*Uncinula*

c) Perithecial appendages dichotomously branched -*Microsphaera*

B. Mycelium partially endophytic

1. Perithecial appendages simple, imperfect state-*Oidiopsis*, *Leveillula*

2. Perithecial appendages coiled at the tip, imperfect state-*Oidiopsis-Pleochaeta*

3. Perithecial appendages with basal swellings, imperfect state –*Ovulariopsis*

II. Ascocarp absent

A. Mycelium superficial

1. Basal cell of the conidiophores swollen.
2. Basal cell of the conidiophores not swollen.

a) Conidia borne in chains–*Euoidium*

b) Conidia borne singly–*Pseudoidium*

B. Mycelium partly endophytic

1. Conidia ovoid, obclavate -*Oidiopsis*
2. Conidia pyriform -*Ovulariopsis*

Order: Clavicipitales

Family: *Clavicipitaceae*

Genus: Claviceps

Order: Meliolales

Family: Meliolaceae

Genus: *Meliola*

Class- Plectomycetes

Order: Eurotiales

Family: Eurotiaceae

Genus: *Aspergillus* and *Penicillium*

Class-Discomycetes

Discomycetes are classified into seven orders. Only Pezizales and Tuberales are discussed here.

Order: Pezizales

Family: Pezizaceae (uninucleate ascospores present)

Genus: *Peziza*

Order: Tuberales

Family: Tuberaceae

Genus: *Tuber*

Subdivision-Basidimycotina

General characters

- Sexual spores are basidiospores.
- Usually four basidiospores are develop on basidium.

Key to the classes of Basidiomycotina

- Basidiocarp lacking and replaced by teliospores grouped in sori or scattered within the host tissues-**Teliomycetes**
- Basidiocarp usually well-developed, Basidia typically organized as a hymenium; Saprobes or rarely parasites :
  - Hymenium present and exposed before the spores mature. Basidiospores are violently discharged-**Hymenomycetes**
  - Basidiocarp remains closed at least until the basidiospores have been released from the basidia. Basidiospores not released with force; Basidium not involved in the discharge of spores-**Gasteromycetes**.

Class: Teliomycetes

- This class includes rusts and smuts.

This class is divided into 2 orders:

1. Uredinales
2. Ustilaginales

Order: Uredinales

Family: Pucciniaceae

- Called rust fungi
- Genus: *Uromyces, Puccinia*

Family: Melampsoraceae

- **Genus**: *Melampsora*

Order: Ustilaginales

Family: Ustilaginaceae

- Called smut fungi
- Genus, *Ustilago, Tilletia, Neovossia*

Family: Tilletiaceae

- Genus: *Tilletia, Neovossia*

Family: Ganodermataceae

- **Genus:** *Ganoderma*

Class Hymenomycetes

Order: Exobasidales

Family: Exobasidiaceae

- Lack the well-define basidiocarps
- **Genus**; *Exobasidium*

Order: Tulasnellales

Family: Ceratobasidiaceae

Genus: Ceratobasidium and Thanatephorus

Order: Aphyllophorales/ Polyporal

Family: Corticiaceae

Genera: *Chondrostereum, Athelia, Ganoderma*

Order: Agaricales

- Include mushrooms, (edible), toadstools (poisonous) and boletes.

Family: Tricholomataceae

- Genus: *Pleurotus,Armillariella, Marasmius,*

Family : *Amanitaceae*

- Genus: *Amanita,Limacella* and *Termitomyces*

Family: Pluteaceae

- Genus: *Volvariella, Pluteus, Chamaeota.*

Family: Agaricaceae

- Genus: *Agaricus, Cystoagaricus* and *Melanophyllium*

Subdivision: Deuteromycotina

General characters

- Deuteromycotina includes the fungi in which the *'perfect stage'(zygote, ascus, basidium) is either lacking* or has not been discovered so far.
- Because of the apparent absence of any perfect stage or sexual phase, these fungi are commonly called *'imperfect fungi'* or technically 'Fungi imperfecti'.

Classification

The sub-divison Deuteromycotina is divided into following three classes:

1. **Blastomycetes:** True mycelium absent or not well-developed, soma is made up of yeast (budding) cells with or without *pseudomycelium.*
2. **Coelomycetes:** Mycelium well-developed, assimilative budding cells absent. Reproduction by conidia borne in *pycnidia or acervuli.*
3. **Hyphomycetes:** True mycelium is present; mycelium is either sterile forms or bear spores on sporophores, which are *never aggregated in pycnidia or acervuli.*

Class: Hyphomycetes

Order

1. **Agonomycetales or Mycelial sterilia**: Conidia absent except for chlamydospores
2. **Moniliales** : Conidia present, Conidiophores are not organized as synnemata or sporodochia- Conidiophores are organized as *synnemata or sporodochia.*
3. **Stilbellales:** *Synnemata* formed
4. **Tuberculariales:** *Sporodochia* formed

Order: Moniliales

Family : Moniliaceae

Genus : *Monilia*

Family: Dematicaeae

Genera: *Alternaria, Bipolaris, Drechslera, Helminthosporium* and *Pyricularia.*

Family : Stilbellaceae

Genus: *Graphium*

Order: Tuberculariales

Family: Tuberculariaceae

Genera. *Fusarium*

Class: Coelomycetes

Order: Melanconiales

Family: Melanconiaceae

- The fructifications are *acervuli.*
- Genera: *Colletotrichum, Cylindrosporium, Melanconium*

Family 3: Stilbellaceae

- Conidia and conidiophores develop in synnemata

Genus : *Graphium*

Order : Sphaeropsidales

Family : Sphaeropsidaceae

Genus : *Macrophomina, Ascochyta*

Order : Agonomycetales or Mycelia sterilia

Genus : *Rhizoctonia* and *Sclerotium.*

Teleomorphic stage of some Important Pathogens

S.No.	Anamorphic stage	Teleomorphic stage
Cleistothecial ascomycetes		
1.	*Penicillium*	*Talaromyces*
2.	*Oidium*	*Erysiphe*
3.	*Paecilomyces*	*Byssochlamys*
4.	*Aspregillus*	*Eurotium*
Perithecial ascomycetes		
5.	*Colletotrichum*	*Glomerella*
6.	*Fusarium*	*Gibberella*
7.	*Trichoderma*	*Hypocrea*
8.	*Verticillium*	*Hypocrea*
9.	*Graphium*	*Ophiostoma*
10.	*Chalara*	*Ceratocystis*
11.	*Acremonium*	*Epichloe*
Loculoascomycetes		
12.	*Alternaria*	*Lewia*
13.	*Cercospora*	*Myosphaerella*
14.	*Septoria*	*Myocsphaerella*
15.	*Phyllosticta*	*Guignardia*
16.	*Stemphylium*	*Pleospora*
17.	*Bipolaris*	*Cochliobolus*
18.	*Drechslera*	*Pyrenophora*
19.	*Exserohilum*	*Setosphaera*
20.	*Curvularia*	*Cochliobolus*
21.	*Sphaeropsis*	*Physalospora*
Apothecial ascomycetes		
22.	*Monilia*	*Monilinia*
23.	*Botrytis*	*Botryotinia*
24.	*Melanconium*	*Greeneria*

25.	*Cylindrosporium*	*Myocospharella*
26.	*Entomosporium*	*Diplocarpon*
Basidiomycetes		
27.	*Rhizoctonia*	*Thanatephorus*
28.	*Sclerotium*	*Aethalium*

Questions

1. Why the genus written with an initial capital letter and species with small letter?
2. Write the standard endings for various taxa of fungi, starting from phylum. Which taxa donot have standard endings?
3. Short notes

a) Write names of five catastrophic diseases of plants

b) Name the kingdom of fungi

4. Elaborate the similarities and differences between the various kingdoms of fungi.
5. Which is the authentic publication on taxonomy of fungi ?
6. Why deuteromycotina is called fungi imperfecti?
7. Write down the differences between rust and smut.
8. Classify order erysiphales with suitable examples.

13

Bacteria and Mollicutes General Morphological Characters Basic Methods of Classification and Reproduction

Bacteria

Definition

Bacteria belong to prokaryota which encompasses organisms with a *primitive type of nucleus* lacking a clearly defined membrane. The bacteria are smaller than fungi and measure about 0.5 to 1.0 x 2.0 to 5.0μ.

General morphological characters

1. A bacterial cell consists of a cell wall and a compound membrane enclosing protoplasm.
2. The cell wall of bacteria is composed of a *peptidoglycan.* It is composed of acetyle-glucosamine and acetyle-muramic acid.
3. The rigid peptidoglycan layer is located between the cytoplasmic membrane and an outer multiple tract layer.
4. The latter layer is composed of *lipoprotein.* It is common in *gram negative* bacteria.
5. Inside the protoplasm, nucleus, vacuoles, mesosomes, polysaccharides, lipids, mitochondric granules and spores are found.
6. Externally bacterial cells may show *capsules, pili, fimbrae* and *flagella.*
7. Flagella are useful for *motility* of the bacteria.The bacteria belonging to *Eubacteriales* have *peritrichous* flagellus at all the sides.
8. Pili serve as adsorption organs for bacteriophages and also mediate *conjugation of bacteria.*

9. Bacteria do not have characteristic nucleus. They contain bodies within the cytoplasm that are regarded as a nuclear structure and DNA is confined to this area.
10. Bacteria contain *70s ribosomes* depending upon the bacteria. Ribosomes are the sites of *protein synthesis.*

Plant Pathogenic Bacteria

General characteristics

1. Most of the plant pathogenic bacteria are *rod shaped* except *Streptomyces* which are mostly *filmentous.*
2. Majority of plant pathogenic bacteria are *aerobic* except; *Erwinia*
3. Plant pathogenic bacteria required optimum growth temperature ranging between 27-30°C.
4. Most of the plant pathogenic bacteria reproduce by asexual process known *as binary fission or fission.*
5. Most of the plant pathogenic bacteria are *gram negative* except *Streptomyces, Clavibactor, Curtobacterium, Nocardia* and *Rhodococcus* which are gram positive in Gram's reaction.
6. In *Streptomyces, Clavibacter* and *Bacillus* a part of cell membrane envaginates into the cytoplasm to form complex membrane infolding called as *mesosomes.*
7. *Bacillus* and *Clostridium* species produce *endosproes* which are dormant structures.
8. Bacterial cell lacking cell wall is called *protoplast.*
9. The naked protoplast may synthesize a portion of cell wall material. Such a protoplast is called *sphaeroplast.*

Basic methods of classification

Bacteria can be classified on the basis of size, shape, structure, and arrangement of cells.

a) Size of Bacteria: Size may varies depending upon the species.

1. Size generally ranges from 1 to 10 microns.
2. Most of the bacterial cells 0.5 to 1 μm in width or diameter.
3. Cylindrical or rod (Baclli) 2-3 μm length.
4. Spiral or helical (Spirlli) – 0.75 -1.25 μm.

b) Shape in Bacteria

The shapes of bacterial cells are

1. Spherical or Ellipsoidal or Oval- **Cocci**
2. Cylindrical or rod- **Bacilli**
3. Spiral or Helical- **Spirilli**

Some species have variety of shapes and thus termed as pleomorphic. E.g. Arthrobacter.

c) Cell grouping or Arrangement in cocci

1. **Monococcus:** Single spherical bacterial cell.
2. **Diplococcus:** A coccus divides into two plane and cells remains in pairs.
3. **Streptococcus:** A coccus is arranged in a long chain e.g. Streptococcus sp.
4. **Tetrad:** A coccus divides in two plans, second division at a right angle to the first plane of division and forms a square of four cells. e.g. *Tetracoccus* sp.
5. **Sarcina:** A cube of eight coccus cell is formed by three divisions in alternate planes at right angle to each other. A cube of eight cells is known as Sarcina.
6. **Staphylococci:** A coccus cells divides in three planes in irregular pattern like cluster of cells or bunch of grapes. E.g. *Staphylococcus albus*
7. **Vibrio:** The short comma shaped cells are called as *Vibrio*. Short tightly coined rods are called *spirillum*. Very long cell with several cuvves and twiste are called "*Spirochete*".

d) Flagellar arrangement in bacteria

All types of bacteria do not have a flagella they are mostly present in bacilli and spirilli and rarely in cocci.

1. **Atrichous:** A cell without flagella. *Ex. Crynbacterium diptherae*
2. **Monotrichous:** Single polar flagellum at one end. *Ex. Vibrio cholerae*
3. **Amphitrichous:** A cell with a bunch of flagellum at both the ends. E.g *Rhodospirillum rubrum*.
4. **Lophotrichous:** A cell having tuft or bunch of flagella at one ends. E.g. *Pseudomonas*.
5. **Peritrichous:** The flagella are evenlly distributed surrounding the entire bacterial cell. *Ex. Bacillus*

Classification of Plant Pathogenic Bacteria

The bacteria belongs to the class Schizomycetes. Plant pathogenic belongs to 3 different orders.

1. Order: Pseudomonadales

Family: Pseudomonadaceae

Important Characters

- It does not produe non water soluble pigment.
- It produces soluble pigment which is not yellow.
- No acid is produced from lactose.

Genera: *Pseudomonas*

Important Characters

- It does not produce non water soluble pigment.
- It produces soluble pigment which is *not yellow*
- *No acid* is produced from lactose.

Important Plant Pathogens

S.No.	Name of Pathogen	Name of disease
1.	*Pseudomonas solanacearum*	Wilts of Tomato, Banana, Brinjal, Potato
2.	*Pseudomonas tabaci*	Wild fire of tobacco
3.	*Pseudomonas syringae pv. phaseolicola*	Halo blight of beans

Genera: *Xanthomonas*

Important Characters

- Culture develop an *yellow non water soluble pigment*
- Produce *acid from lactose*.

Important Plant Pathogens

S.No.	Name of Pathogen	Name of disease
1.	*Xanthomonas campestris*	Black rot of cabbage
2.	*Xanthomonas campestris* pv. *oryzae*	Bacterial blight of rice
3.	*Xanthomonas campestris pv. citri*	Citrus canker
	Xanthomonas campestris pv. malvacearum	Black arm of cotton

2.Order: Eubacteriales

I. Family: Enterobacteriaceae

Important Characters

- The bacterial cells are *gram negative* and motile with *peritrichous flagella.*
- They produce *acid from sugars*.

Genus: *Erwinia*

Important Plant Pathogens

S.No.	Name of Pathogen	Name of disease
1.	*Erwinia amylovora*	Fire blight of apple
2.	*Erwinia carotovora*	Soft rot of many vegetables

II. Family: Rhizobeaceae

Important Characters

- The cells are rod shaped, *gram negative* and sparsely flagellated.
- The colonies are *white.*
- The genus *Agrobacterium* is pathogenic on plants.

Genus: *Agrobacterium*

Important Plant Pathogens

S.No.	Name of Pathogen	Name of disease
1.	*Agrobacterium tumefaciens*	Crown gall in many plants

III. Family: Corynebacteriaceae

Important Characters

- The bacterial cells are *gram positive*.
- *Corynebacterium* is the only gram positive plant pathogen while others are gram negative.
- The bacterial cells are *pleomorphic* rod that show the characteristic arrangement produced by snapping division.
- They are motile rods.

Genus: *Corynebacterium*

Important Plant Pathogens

S.No.	Name of Pathogen	Name of disease
1.	*Corynebacterium michiganense*	Bacterial wilt of tomato
2.	*Corynebacterium tritici*	Tundu disease of wheat
3.	*Corynebacterium sepedonicum*	Leaf spot and wilt in many crops

3. Order: Actinomycetales

Family: Streptomycetaceae

Important Characters

- It produces *conidia in aerial hyphae in chains.*
- It has characteristic branching mycelium and spores are formed by fragmentation of the plasma straight or rod shaped.

Important Plant Pathogens

S.No.	Name of Pathogen	Name of disease
1.	*Streptomyces scabies*	Potato scab

Gram Staining

- Bacterial species are often distinguished from one another by Gram staining.
- In this process, a bacterial smear is heat fixed on glass slide, stained with crystal violet and mordanted with iodine and finally rinsed with ethanol.
- When the bacteria retain the crystal violet stain after rinsing, the bacteria are called gram positive; and those which do not retain the stain are called gram negative.
- The later are then counter stained with pink colour safranin.
- The ability of bacteria to retain crystal violet stain or not, depends upon fundamental structure of cell wall.

Gram positive vs Gram negative bacteria

S. No.	Gram Positive bacteria	Gram Negative bacteria
1.	Cell wall is thicker and homogenous	Cell wall is thinner and usually thin layered.
2.	Contains lower content of lipids (5-10%)	Contains higher content of lipids (up to 40%)
3.	Peptidoglycan comprises up to 90% of the cell wall	Peptidoglycan comprises only 10%.
4.	Techoic acid present	Techoic acid absent
5.	Cell wall has higher amino sugar content (10-20%)	Low content of amino sugars

6.	Cell wall is simple in shape and is single layered.	Varying cell wall shape and is tripartite (3-layered).
7.	Mesosomes more prominent.	Mesosomes less prominent.
8.	Retains violet dye	Retains red dye
9.	Examples: Bacillus, Clavibacter Streptomyces	Examples: Erwinia, Pseudomonas, Xanthomonas, Agrobacterium, Xylella

Reproduction of Bacteria

1. Fission

- The most common mode of reproduction in bacteria.
- The bacterial cells divide into *two daughter cells.*
- It is *asexual* reproduction.

2. Conjugation

- Genetic material of one cell is transferred to another cell during conjugation.
- The two cells are genetically different.
- The donor cell transfers part of its genome to the recipient cell.
- Conjugation is one type of *genetic recombination*
- Genetic recombination refers to *any process leading to the formation of a new individual which derives some of its gene from one parent and some from another, genetically different, parent.*

3. Transforamtion

- DNA from one type of bacterium is incorporated into the genetic makeup of another organism during transformation.
- In this DNA is absorbed through external source.
- This is a method employed in laboratory to bring about *recombination* : but it may be occurring in nature as well.

4. Transduction

- Virus that infects bacteria is called *bacteriophage.*
- Bacteriophage grows within a bacterial cell.
- The infected bacterial cell bursts and the bacteriophage particles are liberated.
- Some of the bacteriophage particles released carry some genetic material of the host bacterial cell.

- They enter into another bacterial cell and the genetic material brought into the cell by the virus is incorporated into that of the bacteria .
- By this process a new genotype of the bacterium arises and this process is called transduction.

Mollicutes

- The mollicutes are a class of bacteria distinguished by the *absence of a cell wall.*
- The word mollicutes is derived from the latin *mollis* (meaning *soft or pliable*) and *cutis* (meaning skin).
- They are parasites of various animals and plants, living on or in the hosts cells.
- Individuals are very small , typically only 0.2-0.3 μm in size and have a very small genome size.
- They vary in form, although most have sterols that make the cell membranes somewhat more rigid.
- Many are able to move about through *gliding*, but members of the genus *Spiroplasma* are helical and move by *twisting* .
- The best known genus in mollicutes is *Mycoplasma.*

Questions

1. Define bacteria? Mention pointwise general characteristic of plant pathogenic bacteria.
2. Classify plant pathogenic bacteria on the basis of size, shape, structure, and arrangement of cell.
3. Write down the procedure of gram staining.
4. Write down the differnce between gram+ve and gram-ve bacteria.
5. Discuss in details about reproduction in bacteria.
6. Write dowm the characters of mollicutes.

14

Viruses: Nature, Structure, Replication and Transmission

Definition of Virus

According to Luria and Darnell (1968), *Viruses are entities whose genome is a nucleic acid, either DNA or RNA, which reproduce inside living cells and use their synthetic machinery to direct the synthesis of specialized particles- the* ***virions,*** *which contain the viral genome and transfer it to other cells.*

Nature of Virus

The physical and chemical nature of viruses eluded scientists until the 1930s. Stanley (1935) announced *crystallization* of Tobacco mosaic viruses, a feat for which he got a Nobel Prize. Stanley thought the crystals to be pure proteins. It was the pioneer work of F.C. Bawden and N.W. Pirie of England which showed that the viruses consists of proetein and nucleic acid. The chemical nature of viruses was settled. It consisted of nucleic acid and protein. The viruses chemically were proved to be nucleic acid. The mature particle of a plant virus is generally called *virion* and the whole infective particle is called as *Nuclcocapsid.*

Nucleic Acid of Viruses

1. Nucleic acid portion of the virus particle is called as *Genome.*
2. Genomes are organized as single nucleic acid molecules that are linear or circular.
3. They may have as few as four genes or as many as several hundred.
4. They may be double-stranded DNA, single-stranded DNA, double-stranded RNA, or single-stranded RNA
5. Genome is actual infective component.
6. Majority of plant viruses contain *RNA*.

7. Some viruses (Cauliflower mosaic virus, Maize streak virus, Mung bean yellow mosaic virus) contain DNA.
8. Most plant viruses contain single strand of RNA.
9. In double stranded viruses (as RNA or ds DNA) the two strands are coiled around each other helically.

The Virus Protein

1. Protein coat that encloses the viral genome is called **capsid.**
2. Its structure may be rod-shaped, polyhedral, or complex.
3. Composed of many **capsomeres,** protein subunits made from only one or a few types of protein.
4. It helps viruses infect their host.
5. It protects nucleic acid (RNA or DNA) of virus.
6. It is made-up of different amino acid sequences in different viruses.
7. Protein shell or coat protects viral nucleic acid from environment.
8. Viruses rely on living host cells for most of the enzymes necessary for their replication.
9. Viruses are in cable of growing for the synthesis of *Lipman system* is absent in viruses.

Structure of Viruses

Plant viruses are usually described as

1. Elongated (Rigid rod or Flexious thread):

S.No.	Morphology	Example	Size(nm)
1.	Rigid rod	*Barley stripe mosaic virus*	20X10
		Tobacco mosaic virus	15 X 300
2.	Flexuous thread	*Potato virus X*	10-13 X480
		Citrus tristeza virus	10-14 X 2000

2. Spherical (Isometric or Polyhedral) : All spherical viruses are actually polyhedral ranging in diameter about 17 nm to 60 nm. Examples;

S.No.	Example	Diameter (nm)
1.	Tobacco necrosis satellite virus	17
2.	Wound tumour virus 60 nm in diameter.	60
3.	Tomato spotted wilt virus 70-80 nm in diameter	70-80

3. Rhabdo Viruses: Bacillus like-These are short bacillus like rods approximately 3 to 5 times as long as they are wide.

S.No.	Example	Diameter (nm)
1.	Potato yellow dwarf virus	75 x 380 nm.
2.	Wheat striate mosaic virus	65 x 270 nm.
3.	Lettuce necrotic yellow virus	52 x 300 nm.

Viral Replication

Viral replication is the formation of biological viruses during the infection process in the target host cells. Viruses must first get into the cell before viral replication can occur. Through the generation of abundant copies of its genome and packaging these copies, the virus continues infecting new hosts. Replication between viruses is greatly varied and depends on the type of genes involved in them. Most DNA viruses assemble in the *nucleus* while most RNA viruses develop solely in *cytoplasm*. Viruses multiply only in living cells. The host cell must provide the energy and synthetic machinery and the low molecular-weight precursors for the synthesis of viral proteins and nucleic acids. The virus replication occurs in seven stages, namely;

1. Adsorption
2. Entry
3. Uncoating
4. Transcription / mRNA production
5. Synthesis of virus components
6. Virion assembly
7. Release (Liberation Stage)

1. Adsorption

The virus attaches to the cell membrane of the host cell. It then injects its DNA or RNA into the host to initiate infection. In animal cells these viruses get into the cell through the process of endocytosis which works through fusing of the virus and fusing of the viral envelope with the cell membrane of the animal cell and in plant cell it enters through the process of pinocytosis which works on pinching of the viruses.

2. Entry

The cell membrane of the host cell invaginates the virus particle, enclosing it in a pinocytotic vacoule. This protects the cell from antibodies like in the case of the HIV virus.

3. Uncoating

Cell enzymes (from lysosomes) strip off the virus protein coat. This releases or renders accessible the virus nucleic acid or genome.

4. Transcription/mRNA production

For some RNA viruses, the infecting RNA produces messenger RNA (mRNA). This is translation of the genome into protein produces. For others with negative stranded RNA and DNA, viruses are produced by transcription then translation. The mRNA is used to instruct the host cell to make virus components. The virus takes advantage of the existing cell structures to replicate itself.

5. Synthesis of virus components

The following components are manufactured by the virus using the host's existing organelles:

- Viral protein: Viral mRNA is translated on cell ribosomes into two types of virus protein.
- Structural: Proteins which make up the virus particle are manufactured and assembled.
- Non-structural: Protein not found in virus particle, mainly enzymes for virus genome replication.
- Viral nucleic acid synthesis (genome replication): New viral genomes are synthesized; templates are either the parental genome or newly formed complementary strands, in the case of single stranded genomes. These genomes are made by either a viral polymerase or (in some DNA viruses) a cellular enzyme, particularly in rapid dividing cells.

6. Virion Assembly

A virion is simply an active or intact virus particle. In this stage, newly synthesized genome (nucleic acid), and proteins are assembled to form new virus particles. This may take place in the cell's nucleus, cytoplasm, or at plasma membrane for most developed viruses. For some RNA viruses, the infecting RNA produces messenger RNA (mRNA). This is translation of the genome into protein produces. For others with negative stranded RNA and DNA, viruses are produced by transcription then translation. The mRNA is used to instruct the host cell to make virus components. The virus takes advantage of the existing cell structures to replicate itself.

7. Release (Liberation Stage)

The viruses, now being mature are released by either sudden rupture of the cell, or gradual extrusion (budding) of enveloped viruses through the cell membrane. The new viruses may invade or attack other cells, or remain dormant in the cell. In the case of bacterial viruses, the release of progeny virions takes place by lysis of the infected bacterium. However, in the case of animal viruses, release usually occurs without cell lysis.

Different types of viruses

The following are the examples for different types of viruses

S.No.	Types		Examples
1.	**RNA viruses**		
	a) Single stranded	With one segment	Tobacco mosaic viruses, Lettuce necrotic yellow virus
		With two segments	Cowpea mosaic virus, Tobacco ring spot virus
		With three segments	Cucumber mosaic virus
		With more number of segments	Potato virus x, Tomato spotted wilt virus
	b) Double stranded		Wound tumour virus, Fiji disease virus, Reo virus
2.	**DNA viruses**		
	Double stranded		Cauliflower mosaic virus
	Single stranded		Maize streak virus, X174

Transmission of Plant Viruses

The plant virus rarely, if ever, come out of plant spontaneously. For this reason, Plant viruses are not disseminated as such by wind or water. Viruses are transmitted from plant to plant in a number of ways such as vegetative, propagation, mechanically through sap and by seed, pollen, insect, mites, nematodes, dodder and fungi.

1. Mechanical Transmissions

Mechanical transmissions may takes place between closely spaced plants after a strong wind by contact, when plants are wounded during cultural operations, virus infected sap adhering to the tools, worker hands or cloth accidently transmitted to the subsequently wounded plants. Example, Potato virus: (PV-X, TMV on Tobacco and Tomato).

2. Transmission by Vegetative Propagation

Plants are propagated vegetatively by budding or grafting or by cutting or by the use of tubers, corms, bulbs or rhizome. Any virus present in the mother plant from which these organs are taken will almost always be transmitted to the progeny. Transmission of viruses may also occur through natural root grafts of adjacent plant. Example, Mosaic and leaf curl of respberry.

3. Transmission by Seed and Pollen

Majority of the seed transmitted viruses are carried internally. The virus may arrive to the seed from the infected ovule or pollen. Of the large number of virus disease of plants, a small percentage show the seed transmission. The common examples are muskmelon mosaic virus in watermelon, barley stripe mosaic virus in barley, tobacco ring spot virus in soybean and common bean mosaic virus.

4. Dodder Transmission

Several plant virus can be transmitted from one plant to another plant through the bridge formed between the two plants by twining the stem of parasitic plant, dodder (Cuscuta species), green strain of cucumber mosaic virus on *N. glutinosa*, lucern (alfa-alfa), mosaic virus to tobacco, potato stem mottle virus to tobacco. Bennett (1940) showed dodder transmission, sugar beet curly top virus (BCTV), cucumber mosaic virus.

5. Natural Modes of Transmission

This includes air borne transmission through insects and mites and soil borne transmission through nematodes and fungus.

A) Air Borne through Insects

The most common and important means of virus transmission in the field. Members of the order homoptera like aphids, jassids, leaf hopper, white flies, mealy bug, scale insects and thysanoptera like thrips, coleopteran-beetles play major role in transmission of viruses. Insects with sucking mouth parts carry plant viruses on their stylet- stylet borne or non persistances.

B) Circulative

1. Circulative Viruses (Persistent): A circulative virus is one that passes into a vector through the mouth parts, circulates internally and enlarges through the salivary glands.
2. Propagative Viruses: Propagative viruses are those viruses which multiply in their insect vectors and transmit them for a long time but very often for as long as they live.

C) Insect Vectors

Virus transmitting insect is called *vector*. Viruses are mostly really on insect for transmission (400 species of insect vector transmitting more than 200 viruses). The viruses are transmitted by diffeent types of insects and most efficient vectors are of sucking and bitting type of insects.

a) **White Flies:** Yellow vein of okra, Tobacco leaf curl, Pumpin yellow mosaic, Mung yellow mosaic, Sweet potato mosaic, Cotton leaf curl, Cassava mosaic.

b) **Aphids:** These are the most important insect vector of plant viruses and transmit great majority (about 170) of the all stylet borne viruses Example. Soybean mosaic, Pea enation mosaic, Lettuce necrotic, Yellows borne mosaic, Potato leaf roll, Red clover mosaic, MCMV, AMV, BYMV.

c) **Leaf Hoppers:** All leaf hoppers transmitted viruses are circulatory, several are known to multiply in the vector (propagative), cause disturbance in phloem region Example, Rice tungro viruses, Rice dwarf viruses, Potato yellow dwarf, Maize mosaic, Beet curly top, Maize rough dwarf.

d) **Thrips:** Tomato spotted wilt virus.

e) **Beetles:** Cowpea mosaic virus, Bean pod mottle virus, Squash mosaic, Cowpea chlorotic motile virus, Raddish mosaic.

f) **Mites:** Transmit viruses like Pigeonpea sterility mosaic, Peach mosaic, Fig mosaic, Wheat streak mosaic.

D. Nematodes

Hewitt and colleagues (1958)first showed that *Fan leaf virus of grapevines* is transmitted by dagger nematode, *Xiphinema index*. Later, a few other plant viruses were reported to be transmitted by nematodes and now about 24 plant viruses are known to be transmitted by neamtodes. All species of nematodes known to transmit viruses are members of the order *Dorylaimida* and belong to the five genera: *Xiphineama, Longidorus, Paralongidorus, Trichodorus* and *Paratrichodorus.* Examples of viruses that can be transmitted by nematodes are given below.

S.No.	Group	Nematode Virus Vectors	Example
1	NEPO: Nematode transmitted polyhedral viruses	*Xiphinema index* *Xiphinema coxy* *Xiphinema americaum*	Grape vine fan leaf Cherry leaf roll Ring spot of tobacco tomato, Soybean bud blight, Bringal mosaic.
2	NETU: Nematode transmitted tubular viruses	*Longidorus*	Rasberry ring spot, Tomato black ring.
		Trichodorus and Para trichodorous	Rod shaped viruses, Tobacco rattle, Pea early browing.

E. Fungus

Soil inhabiting speices of chytrid true fungi belonging to two genera *Olpidium* and *Synchitrium* and 2 species of protozoa (fungal like) belonging to two genera *Polymyxa* and *Spongospora* have been proved to be vectors of certain category of plant viruses. Examples of viruses that can be transmitted by fungi are given below.

S.No.	Fungus Virus Vectors	Example
1.	*Olpidium brassicae*	Lettuce big vein, Tobacco necrosis, Cucumber necrosis, Tobacco stunt virus
2.	*Synchytrium endobioticum*	PU-X virus
3.	*Polymyxa graminis*	Wheat mosaic virus, Beat necrotic yellow vein virus.
4	*Spongospora subteranea*	Potato mop top virus

Terminology

1. **Satellite viruses:** These are viruses associated with certain typical viruses but depend on the latter for multiplication and plant infection and reduce the ability of the typical viruses act like parasite of the associated typical virus.
2. **Helper viruses:** The viruses which help multiplication of satellite RNA are called helper viruses.
3. **Virusoids:** These are viriod like, small, single stranded , circular RNAs that are present inside some RNA viruses, form an obligatory association with these viruses so that neither the virus not the virusoids multiply and infect a plant in the absence of its partner.
4. **Temperate viruses:** Viruses that can integrate their genome into a host chromosome and remain latent until they initiate a lytic cycle.
5. **Provirus:** Viral DNA that inserts into a host cell chromosome.
6. **Retrovirus:** RNA virus that uses reverse transcriptase to transcribe DNA from the viral RNA genome.
7. **Viroid:** These are small (250-400 nucleotide), naked, single stranded, circular RNAs capable of causing disease in plants by themselves.
8. **Transovarial transmission:** Viral transmission from an insect vector to its offspring, meaning off spring of an infected vector are also infective.

Questions

1. Define Virus, Viroid & Prions
2. Write notes on any of the four i. Gemini viruses ii. Major group of viruses iii. Segmented viral genomes iv. dsDNA plant viruses.
3. Differentiate between i. Viruses and Viroids ii. Satellite virus and Satellite nucleic acids iii. Viroids and Prions.
4. Define: Provirus, Retrovirus, Transovarial transmission, Temperate virus, Satellite virus.
5. Describe transmission of plant viruses?

6. Explain virus vector relationship.
7. Classify plant viruses on the basis of nucleic acid with suitable examples.
8. Write an essay on management of plant viruses.
9. Describe the transmission of plant viruses by any four of the following

i) Aphids
ii) White flies
iii) Thrips
iv) Beetles
v) Planthoppers

15

Phanerogamic Parasites

The parasitic flowering plants are also important pathogens. There are few seed plants, which are parasitic on living plants and are called parasitic higher plants or *Phanerogamic parasites*. These parasitic higher plants attack some valuable crops and trees causing considerable losses. They produce flowers and seeds and belongs to several widely separated botanical families they differ to each other on their dependency on host plants. These parasites have *haustoria* as *absorbing organ,* which sent deep into the vascular bundle of the host to draw water and nutrients. More than 2500 species of higher plants are known to live parasitically on other plants.

Classification of flowering plant parasites

i) Complete Parasite (Holoparasites)

a) Root: Orobanchea (Broom rape)

b) Stem: Cuscuta (Dodder Amarvel)

ii) Partial Parasites (Semi Parasites)

a) Root: Striga (Sandle wood witch weed)

b) Stem: Loranthus (Banda or Deudrophae)

Complete Root Parasite

Orbanche (Broom Rape)

1. Orobanche spp. are *total root parasites* affecting tobacco, brinjal, tomato, cauliflower, turnip and many other solanaceous and cruciferous plants.
2. The parasite consists of a stout, fleshy stem 15 to 50 cm long. This stem is yellow or brownish red in colour and is covered by small thin and brown scaly leaves.
3. The flower appearing in the axil of the leaves are white and tubular seeds are very small and black in colour and many remain viable in soil for several years.

4. The *haustoria* of the parasite penetrate into the root of the hosts and draw it nourishment.
5. When the host is carefully uprooted, the parasitic roots are seen intertwined with the host root systems; the growth of the host is retarded and remains stunted.

Management

1. Long crop rotation.
2. Destroy the parasite before flowering.
3. Drenching of the soil with 0.25% copper sulphate.
4. Fumigating the soil with methyl bromide.
5. Treat with the herbicides "Glyphosate".

Complete Stem Parasite

Dodder (Cuscuta)

1. Non-chlorophyll bearing leafless, twining parasitic seed plants, they attach to the host.
2. Yellow, pink or orange in colour.
3. When the stem of parasites comes in contact with the hosts, the minute root like organs, *haustoria* penetrates into the host cortex and serves as an *organ of food absorption.*
4. When the relationship with the host is firmly established, the dodder plant losses the contact from soil.
5. The seeds are formed in *capsule.* A single plant may produce as many as 3000 seeds.
6. The common dodder *Cuscuta granovil* attacks *clovers, berseem flag* and many other *oilseed crops*. It also attacks *ornamental* and *hedge plants*.
7. It is fast developing parasites and within 2-3 seasons may destroy a complete plant.
8. The dodder perpetuates through *seeds* which remain dormant in the soil until favorable seasons returns, stem portion of parasite is also meant of perpetuation.

Management

1. Selection of dodder free seeds for sowing.

2. Crop rotation with non host plants.
3. Restriction of flow of irrigation water through infected field.
4. Use of soil herbicides, such as dichobhil, donoseb, DCPA, glyphosate, these chemicals kills the dodder plant upon its germination from seed.
5. Preventing the movement of grazing animals from infected field to clean field.

Partial Root Parasite

Striga (Witch weed)

1. Striga is *partial root parasite* of *sugarcane, cereals, jawar, maize* and *millets* in India.
2. There are four species of Striga reports in the country on sugarcane, rice, sorghum and other millets like *S.lutee S.densiflora, S. quphrasioldes, S.asiatica.*
3. These plants although *obligate parasites* do not obtain all of their nutrient material from their host root. They posses' *chlorophyll bearing leaves.*
1. One flower / capsule contains 1200-1500 seeds. Short distance dissemination occurs through seeds.
2. Stem of parasite forms *haustoria* which penetrates the root of hosts plants and also water and nutrient eventually wasting and destroying the host.

Management

1. Deep ploughing after harvest.
2. Complete eradication of parasite before flowering.
3. Regular intercultural should be followed.
4. Crop rotation with cotton-jowar- groundnut.
5. Weedicides are used to control Striga before flowering.

- 2-4-D @ 2.5 lit/500 lit of water O per ha.
- Attrazine @ 2 kg/500 lit of water per ha.
- 1% TCPA @ 45 kg/ha.

Partial Stem Parasite

Loranthus or Bandgul

1. Loranthus is common parasites of mango trees.

2. The parasite attacks, the aerial parts of the host trees, by developing *haustoria* and obtain its nourishment directly from the vascular system of the host plant.
3. Since the parasite attacks the aerial part of host tree, situated far above the soil level and as such devoid of root system of its own.
4. The fruit is fleshy and contains a solitary seed. It is sweet eaten by birds and animals.
5. The parasite is spread by dispersal of its seed, mostly through birds and to some extent by other animals.
6. The damage done by the parasite is most marked in production of new growth of the host. Leaves are reduced in size and show unhealthy green colour. The quantity and yield of fruit is considerably lowered.

Management

1. Injection of $CuSO_4$ or 2-4-D into the infected branches.
2. Scrapping of parasite before seedling from the infected bunches.
3. Sowing of branches sufficient low to the tumours.

Questions

1. Describe the lifecycle and management of orbanchae?
2. Which are the four most damaging parasitic plant genera?
3. Which are the common stem parasitic angiosperms?
4. What are thee types of mistletoes?
5. Why is family scaceae called Christmas mistletoe family?
6. How is ethylene used to control stirga?
7. Distinguish between.

I. Holoparasitic and Hemiparasitic angiosperms.

II. Orbanchae and Striga.

III. Host range and host preference of parasitic angiosperm.

16

Plant Parasitic Nematodes General Morphology, Reproduction Classification, Symptoms and Nature of Damage

Definition

Neamtode may be defined as triploblastic, bilaterally symmetrical, unsegmented, pseudocoelomate individual invertebrate, with 4 main hypodermal chords, a triradiate oesophagus, circum oesophageal nerve ring, and lacking specialized organs for respiration and circulation.They have tubular gonads which open separately in females but join alimentary canal in males to open through a cloacal aperture.

General Morphology

1. Plant nematodes are tiny worms, size commonly ranges from 0.5 to 1.5 mm, thread like, cylindrical, tapering toward the head and tail.
2. They are characterized by the possession of a needle like *stylet* or spear at their anterior end which is used to puncture the plant and ingest the cell contents.
3. The body is covered with by a transparent cuticle ,which bears surface marks helpful for identifying nematode species.
4. The body structure of nematodes consists of a flexible body wall which is made up of *cuticle, hypodermis* and *somatic muscles*.
5. The wall surrounds a tube like gut which is made up of *oesophagus and intestine*.
6. The body cavity between the gut and the body wall is usually regarded as a *pseudocoelom* containing a pseudocoelomic fluid.
7. Plant parasitic nematodes possess all of the major organ systems of higher animals except respiratory and circulatory system.
8. The life cycle of a plant parasitic nematode has six stages; egg, four juvenile stages and adult.

9. Nematodes belong to the animal kingdom, and phylum *Nematoda.* Most of the important parasitic genera belong to the order *Tylenchida* and few under Dorylaimida.

Reproduction

Nematodes are *dioecious* or *unisexua*l, i.e. male and females are separated. In general appearance both the sexes look alike and are similar in respect of other body systems. However, males can be distinguished by the possession of *spicules* and usually *bursa* near the posterior end. In both male and females the reproductive system is again tubular, called genital tract, which is suspended in body cavity.

1. Female Reproductive system

- The various part of the female reproductive system are: *ovary, oviduct, spermatheca, uterus, vagina* and *vulva*.
- The oocytes are formed in the ovary where they grow in size, pass down to spermatheca via oviduct.
- Spermatheca, if present stores the sperms after mating and when the oocytes pass through it, they get fertilized.
- Further layers on the egg (ovum) are laid while it is passing through the grandular portion of the uterus.
- Vagina helps in the expulsion of egg to the outside and opens on the ventral side through a slit like opening, the vulva.
- Females may possess only *one gonoduct* (*monodelphic*) which may be directed *anteriorly (prodelphic)* or *posteriorly (opisthodelphic*).
- Female posses *two gonoducts (didelphic),* the uteri of two gonoducts unite to form a common vagina leading to vulva.
- In didelphic forms, one gonoduct may be directed anteriorly and the other posteriorly *(amphidelphic),* or both may be directed anteriorly *(prodelphic),* or posteriorly *(ophisthodelphic*).

2. Male Reproductive system

- Among plant parasitic nematodes, male generally have a *single* genital tract (monarchic).
- Various parts of male reproductive system are: *testis, seminal vesicle, vas deferens, cloaca* and *cloacal aperture*.
- Sperms are formed in the testis, and may be temporarily stored in the seminal vesicle.

- Vas deferences joins posteriorily with rectum to form cloaca.
- A pair of hard curved structures, the *spicules* are located in a pouch of cloaca.
- The spicules are protrusible and help in the insemination of females during mating.
- The movement of spicules is guided by a plate like structure, the gubernaculums.
- Bursa is usually present in plant parasitic nematodes. It helps in *clasping* the female during mating.
- Spicules, gubernaculums and bursa constitute secondary sense organs.

Classification

Phytonematode can be classified by the basis of parasitism into three main groups:

1. Ectoparasites
2. Semi endoparasite
3. Endoparasite

1. Ectoparasites

- Feed externally, on the surface of the plant, usually on root hairs , but do not enter inside the plant tissues.
- Only stylet is inserted into the plant cells to draw the nutrition.
- Spend their entire life cycle in soil and move about freely in the rhizosphere.
- Egg are laid in soil.

Ectoparasites can be divided into two groups

a) Migratory ectoparasite

b) Sedentary ectoparasite

a) **Migratory ectoparasite:** When an ectoparasite feed at a particular site for a brief period and the move to the next site. It is called migratory ectoparasite.

Examples: Dagger nematodes - *Xiphinema*

Needle nematodes - *Longidorus*

Stunt nematodes - *Trichodorus* and *Paratrichodorus*

Sting nematode - *Belonolaimus*

b) **Sedentary ectoparasite:** Some ectoparasitic nematode are sluggish, and feed at a particular site for long time.They may secrete a plug to fix their stylet at a particular feeding point. Examples,

Sheath nematode: *Hemicycliophora arenaria*

Sessile nematode: *Cacopaurus pestis*

2. Endoparasites

- Considered to be the most destructive forms of plant parasitic nematodes.
- After infection, they are most completely embedded within the roots, where they feed and reproduce.

These can also be divided into two groups.

a) Migratory endoparasites

b) Sedentary endoparasites

a) Migratory endoparasites

- Mobile nematodes that feed inside the plant root tissue.
- All life stages of migratory endoparasitic nematodes are mobile except the egg.
- Nematodes burrow through the plant from cell to cell, or may leave the plant tissue in search of new feeding sites.

Examples: Lesion nematode- *Pratylenchus*

Burrowing nematode- *Radopholus similis*

Root rot nematode- *Hirschmanniella*

b) Sedentary endoparasites

- Nematodes that, once they have reached a feeding site inside the plant, cease to be mobile and feed from a fixed location.
- Considered to be the most advanced plant parasitic nematodes.
- Female sedentary endoparasitic nematodes generally produce a large number of eggs, which remain in their bodies.

Examples: Cyst nematodes-*Heterodera, Globodera*

Root-knot nematodes-*Meloidogyne*

3. Semi-endoparasite

- Some nematodes, which become only partly embedded in the root tissue.
- Usually, the anterior portion of the body is inserted inside the plant itssues.

- These are again of two types.

  Migaratory semi-endoparasite

  Sedentary semi-endoparasite

a) Migaratory semi-endoparasite

- They donot establish a permanent feeding site and move from one site to another frequently.
- While feeding, the anterior portion of the body is embedded inside the roots to varying degrees, rarely they may be completely embedded inside the plant tissues.

  Examples: Spiral nematode-*Helicotylenchus*

  Lance nematode-*Hoplolaimus*

  Stunt nematode-*Tylenchorhynchus*

b) Sedentary semi-endoparasite

- Establish a permanent feeding site and do not move after infection.
- Posterior portion of the female body outside the roots becomes swollen to varying degrees.

  Examples: Reniform nematode- *Rotylenchulus reniformis*

  Citrus nematode- *Tylenchulus semipenetrans*

Symptoms on crop plants

Below ground plant parts

- Bushy roots.
- Brown lesions spots or speaks on the roots, tuber etc.
- Swelling or typical root galls or knot on the roots.
- Dirty brown colored roots having tendency for adhering of soil particles.
- Forking of roots especially in carrot etc.

On above the ground plant parts

- Stunted growth.
- Yellowing of leaves.
- Less till ring/proliferation of crops.
- Reduced size and number of ear heads, flower, fruits and grains etc.
- Crinkling, curling and twisting of leaves.

- Malformed small and swollen ear heads with diseased grains.
- Trickling of yellow bacterial mass from the leaves stem and ear heads etc.

Nature of damage

The direct or indirect damage to plants by nematodes is caused in several ways.

Indirect damage

- Mechanical wounded agents.
- Host modifiers.
- Rhizosphere modifiers.
- Resistance breakers.
- Vector of pathogens.
- Interference in nitrogen fixation.

Direct damage

- Mechanical injury to the cells.
- Pre emergence death of seedlings.
- Induce feeding areas like giant cells, syncytia etc in vascular tissues, disrupt the flow of water and nutrients from roots to shoots, resulting in poor plant growth.
- Cause abnormalities like root and seed gall in plant due to hormonal imbalance.
- Reduced yield.

Some plant parasitic nematodes with their diagnostic symptom

S. No.	Name of disease	Name of nematode	Diagnostic symptoms
1.	Root knot	*Meloidogyne*	• Galling of roots &tubers • Chlorosis • Wilting
2.	Cyst	*Heterodera* and *Globodera*	• Cysts in roots • Poor root growth • Chlorosis • Wilting
3.	Burrowing	*Radopholus similis*	• Necrosis of roots • Tubers rots • Root breakage • Toppling
4.	Seed gall	*Anguina tritici*	• Glling of seeds • Basal swelling of stem • Ear head sorter and broader

Questions

1. What features are used for identification of plant parasitic nematodes?
2. The nematodes placed in which two class and order?
3. How are namatodes extracted from suspected soils and plant material for identifications?
4. Describe any three modes of feeding in plant nematodes?
5. Write notes on:

i) Classification of plant parasitic nematodes.

ii) Under and above ground symptoms produced by nematodes.

iii) Integrated nematode management.

iv) Reproductive system.

v) Life cycle of nematode.

6. Classify plant parasitic nematodes on the basis of their parasitism?
7. White down the symptoms and nature of damage caused by Heterodera, Meloidogyne and Anguina.

17

Dispersal of Pathogens

Definition

To make a healthy plant sick or diseased, primary requirement of a pathogen is spread of its inoculums from the source of survival to the susceptible host. *The spread of a plant pathogen within the common area in which it is already established is called dispersal or dissemination.* Moving the inoculum only a few inches and transporting it for hundreds of kilometers both represent its dispersal. Nevertheless, the pathogen dispersal is not necessarily only for spread of diseases but also for continuity of the life cycle and development of the pathogen. A detail knowledge of pathogen dispersal is necessary to find out effective control measures for diseases because the possibilities of preventing dispersal and thereby breaking the infection chain always exist.

Plant pathogen dispersal generally occurs through two key groups:

A. **Direct (Autonomous) dispersal:** The pathogen is carried through

1. Soil
2. Seed
3. Planting material

B. **Indirect (Passive) dispersal:** The pathogen spreading itself or by natural agencies like

1. Wind
2. Water
3. Animals
4. Insects
5. Mites
6. Nematodes
7. Birds

A. Direct dispersal

1. **Through seed**: Infected/infested seed is one of the most important means of autonomous dispersal. Dispersal of inoculums through seeds is accomplished as

- Mixture and contaminant dormant structure of pathogens like galls, sclerotia, cysts and smut sori. Example. grain smut of jowar is an example of the former type while the fungal structures called 'sclerotia' Ergot disease of bajra
- Through presence of propagules on the seed coat. Example. Covered smut of barley.
- As dormant mycelium in the seed. Example. Loose smut of wheat, Ring rot and Brown rot of potato, Whip smut and Red rot of sugancane, Mosaic and leaf roll of potato.

2. **Through soil:** A large number of plant pathogens survive in soil and their dispersal takes place in two ways

- Many pathogens survive in soil as facultative parasites or facultative saprophytes. Some of the fungi produce spores which are motile (Zoospores) and can move some distances in soil and infect their hosts. Nematodes can move some distances in soil infect their hosts.
- The contaminated soil is transported to other places along with planting material, seedlings or implements etc. while adopting agronomic practices.

3. **Through vegetative planting materials:** Dispersal of large number of plant pathogens takes place through the vegetative parts of the plants used as planting materials, such as tubers, cuttings, runners, rhizomes, grafts, etc. Over 40% of the bacterial plant pathogens are transmitted on vegetatively propagated materials. Among the important examples are *X. compestris pv.citri* (citrus canker) and *A. tumifaciens* (crown gall).Viruses are invariably transmitted by vegetative propagules; thus virus diseases are mainly severe in vegetative propagated horticultural crops.

Table Autonomous dispersal of plant pathogens.

Autonomous dispersal	Examples
Through seed	
Seed contamination	*Anguina tritici* (seed gall nematode), *Tolyposporium penicillariae* (Smut of bajra), *Claviceps*, *Cuscuta*
Internally seed borne	Loose smut of wheat *(Ustilago tritici)*
Externally seed borne	Covered smut of barley (*Ustilago hordei*), Stinging smut of wheat (*T.caries* and *T. foetida)*, .Black arm of cotton *(Xanthomonas campestris pv. mavacearum)*

Through vegetative planting materials	
Seed tubers	Late blight of potato (*Phytophthora infestans*)
Sugarcane Sets/cutting	Red rot of sugarcane (*Colletotrichum falcatum*)
Propagative materials	Citrus *Tristeza* and greening
Rhizome	Rhizome rot of ginger

B. Indirect dispersal

1. **Wind dispersal:** Fungal spores produced externally on host surfaces are most easily carried by wind currents. Extensive and severe epidemics of plant diseases are mostly the results of wind dispersal of the pathogens. Wind dispersal involves four stages relating to the spores *viz.* production of countless spores, their liberation in the wind currents, dispersal along with the wind and deposition on new susceptible host surfaces where they cause infection under favourable climatic conditions. This is the most dangerous mode of transmission of plant pathogenic fungi like those causing powdery and downy mildews, leaf spots, blasts, blights and rust diseases.
2. **Water dispersal:** Disease dispersal through the agency of water in different ways is comparatively less important as compared to the wind dispersal. Splashing rain drops generally transmit the foliar diseases from leaf to leaf, from shoot to shoot and even from plant to plant in case of closely spaced crops. Plant pathogens requiring high humidity conditions like the fungi causing downy mildew diseases or bacteria causing canker of citrus are well adapted to this kind of short distance water dispersal. Certain soil inhabiting pathogenic fungi and bacteria causing root and collar rots, wilts, foot rots, etc are likely to be transmitted to much longer distances through the agencies like irrigation water, streams and rivers, etc. It is also an important agency in transmission of seeds of higher flowering parasites like dodder and striga.
3. **Insects dispersal:** Most of the viral diseases of plants are transmitted through the agency of different insects. Both types of insects *viz.* sucking and chewing or/biting are capable of transmitting viral diseases. The transmission may be simply 'mechanical' or it may be 'biological'. In case of mechanical transmission the pathogen is simply carried externally or internally by the insect. In the biological case the specific insect and the specific viral pathogen have some kind of association or relationship between the two. Insects in such cases are called the 'vectors' for the particular viral pathogen. Viruses carried 'biologically' by the insect vectors are of two types:

1. **Non-persistent**-viral pathogen requiring no latent or incubation period in the insect body.

2. **Persistent:** Viral pathogens requiring certain incubation period inside the vector body before they are inoculated or transmitted to healthy host. The insects responsible for transmission of viral diseases belong to the species of aphids, jassids (leaf hoppers), white flies, mealy bugs, etc. Certain bacterial *viz*, Fire blight of potato (*E. amylovora*), Citrus canker (*X. citri*) and several fungal pathogens viz, Dutch elm (*Ceratostomella ulmi*), Anthracnose of cucumber (*Colletotrichum lagenarium*) are also known to be carried by insects.

4. **Mites:** The mites are the only non insect arthropod vectors of plant viruses. The ability to transmit virus is restricted to a few species of eriophyd mites. Diseases transmitted by eryophid mites include fig mosic, sterility mosaic of pigeon pea, wheat streak mosaic and wheat spot mosaic.

5. **Nematodes:** Nematodes have been observed to transmit viral, bacterial and fungal plant diseases. Nematodes feeding externally on host plant roots cause injuries to roots which become the avenues for entrance of fungal and bacterial pathogens infecting plant roots. Members of genera *Longidorus* and *Xiphinem*a transmit polyhedral nepoviruses such as the arabis mosaic grapevine fan leaf, tomato ring spot. *Trichodorus* and *Paratrichodorus* transmit the netuviruses such as tobacco rattle and pea early browning.

6. **Animals:** Farm animals serve as disease transmitting agents in some cases. They are likely to carry the pathogen externally on their body surface, particularly on legs and hoofs, etc. or internally through their intestinal tract. Commonly, the soil inhabiting fungi causing rots and wilts are carried externally while certain smut fungi causing diseases to grain crops are transmitted through the intestinal tract.

7. **Birds:** Although birds play a very minor role in disease transmission, in cases of dispersal of seeds of higher flowering parasite. Loranthus sp. parasitising certain trees like mango, etc. their role is of great significance. They transmit loranthus both externally and internally.

8. **Implements and Tools:** Farm implements used for cultivation of soil are often likely to transmit plant pathogens from one place to another. The pathogens in this case are usually carried in the form of bits of plant disease debris lying in the soil. Similarly tools used for carrying out operations like cutting, pruning, budding, grafting, thinning,etc. also help in the transmission of certain diseases from plant to plant. Several viral diseases are disseminated through the budding and grafting operations.

9. **Human dispersal**: Man is often responsible for transmission of plant diseases in two ways viz.

1. Workers handling seedlings, other planting material or fruits are likely to get personally in contact with plant pathogens like fungi or bacteria. While

handling the diseased material and unknowingly and indirectly transmit the pathogens to healthy seedlings or plant parts through his contaminated hands. This is a kind of 'continuous' mode of transmission.

2. The other or'discontinous' mode of transmission for which only man is responsible is the most efficient and equally dangerous phenomenon of transmission of plant diseases between distant geographical areas often separated by physical barriers like oceans, mountains or deserts, etc. Such long distances transmission of a disease to an area or country hitherto free from the disease is usually accomplished by the transport of infected seed, nursery stock or timber, etc. Thus it is a kind of direct transmission through propagating material.

Table : Passive (indirect) dispersal of plant pathogens.

Passive dispersal		**Examples**
Through water	Water splash	Bacterial leaf blight of rice *(Xanthomonas campestris pv. oryzaecola),* Bacterial streak disease of paddy *(Xanthomonas oryzecola)* Coffee rust *(Hemileia vastatrix)*
	Water flow dispersal	Club root of cabbage (*Plasmodiophora brassicae*) Late blight of potato (*Phytophthora infstans*), Red rot of sugarcane (*Colletotrichum falcatum*)
Through Air		Scab of apple (*Venturia inaequalis*), Black stem rust of wheat (*Puccinnia graminisa tritici*), Late blight of potato (*Phytophthora infestans*)
Through insects	Bees and Wasp	Fire blight of apple (*Erwinia amylovora*)
	Beetle (*Scolytus)*	Dutch elm disease (*Ceratocystis elmi*)
	Leaf minor	Citrus canker (*Xanthomonas citri*)
Through nematodes	*Anguina tritici*	Yellow ear rot of wheat *(Corynebacterium tritici*)
	Aphlenchoidea	Leafy gall (*Corynebacterium tritici)*
	Xiphinema index	Grapevine fan leaf virus
Through Fungi	*Synchitrium endobioticum*	Potato virus X, Potato mop top virus
	Olpidium brassicae	Tobacco necrosis virus, Letuce big vein virus
Through birds		Loranthus (*Dendrophthoe*), Mistletoe (*Viscum*), Chestnut blight (*Endothea parasitica*)
Through man		Wart of potato (*Synchitrium endobioticum*), Flag smut of wheat (*Urocystis tritici*)

Questions

1. What do you understand by dispersal of plant pathogens? Discuss autonomous means of dispersal.

2. Name two fungi that help survival and transmission of viral pathogens.
3. Differentiate between primary and secondary inoculums.
4. Name three non-fungal disease that has been introduced in our country from England.
5. Write notes on:

a) Role of insects in pathogen perennation.

b) Dissemination of pathogen inside soil.

c) Anthropochory

6. Give a concise account of pathogen dispersal.

18

Survival of Plant Pathogens

Definition

The availability of a plant pathogens in given area depends upon its ability to survive not only during its parasitic relations with hosts but also during those seasons in which the hosts are not growing. The latter part of *continued existence of plant pathogens in which they remain in dormant condition to overcome the unfavourable condition of non availability of hosts is called "survival*". Pathogens may survive between crop seasons by means of specialized resting structures , by functioning as saprophyte in or on the soil or in diseased plant debris, or by living in some intimate association with living plants or other organism.

Modes of Survival: The modes of survival of plant pathogens are described below:

1. Parasitic survival
2. Saprophytic survival
3. Survival through dormant structures
4. Survival through seeds
5. Survival through insects
6. Survival through nematodes
7. Survival through fungi

1. Parasitic survival

Organisms, parasitic on perennial plants (fruits, plantation and forest trees) survive in or on their hosts actively or passively depending on the environmental conditions. Certain bacterial pathogens *viz*, *X. compestris pv citri, X. compestris pv jugalandis, X. compestris pv. pruni* and *Erwinia amylovora* are known to be carried over perennially in holdover cankers or blighted twigs, which may produce bacterial ooze in favourable weather to fresh inoculums. Similarly several fungal plant pathogens like *Colletotrichum gloeosporioides* (anthracnose of mango), *Oidium mangiferae* (powdery mildew of mango), *Podosphaera leucotricha* (powdery mildew of apple), etc survive on infected organs of the plants. Along with cultivated crops, several undesirable plant (weeds, etc) both annual and perennial, grow independently in the nature. Such plants which belong to the same botanical family to which cultivated hosts belong, are known as *collateral hosts*. Some important examples include the survival of three Puccinia spp. causing wheat rusts in Indian subcontinent. These fungi survive in active sporulating stage on wild species of *Triticum, Aegilops, Hordeum* and *Agropyron repens*. Similarly, rice tungro virus survive actively on *Oryza* spp., *Echinocloa* spp. and *Leersia hexandra.*

Heterocious rust fungi require two kinds of host plants to complete their life cycle. The other host does not belong to the botanical family of cultivated host, is referred to as *alternate host*. Fungal pathogens *Rhizoctonia solani, Sclerotium rolfsii* and *Sclerotinia sclerotium* are known to have wide host range (plants belonging to different families).

2. **Saprophytic survival :** Many plant pathogens adopt survival ability in the absence of their hosts. They grow saprophytically either in or on the soil, or in dead plant debris during this phase. Pathogens that can survive in or on the soil by growing sapropytically are generally considered to be of two types: *soil inhabitants* and *soil invaders. Soil inhabitants are those facultative parasites that are able to survive indefinitely as saprophytes in the soil, and have a wide host range. Pythium, Rhizoctonia* and *Fusarium* are good example of soil inhabitant fungal pathogens. *Soil invaders (soil transients) are those that are rather more specialized parasites generally living in close association with their hosts but may survive in soil for relatively short periods of time as saprophytes.* Most of the bacteria and fungal pathogens of facultative saprophytic nature represent soil invaders.

3. **Survival through dormant structures**: Dormancy is the reversible interruption of phenotype development of an organism. *Dormancy is an adaptation on part of biological organisms to ensure their survival.* Dormant structures, due to less nutrient requirements and hardy structures, have better chance of survival under unfavourable conditions. Fungi, nematodes and phanerogams survives through their resting or dormant structures; phanerogams produce seeds, which can live in dormant stage, sometimes for years. Majority of phytopathogenic nematodes, survive through their dormant structures (cysts, galls, eggs). Plant parasitic fungi produce a variety of resting/dormant structures such as thickened hyphae, chlamydospores, sclerotial bodies, oospores, perithecia, cleistotheica etc. Plant pathogenic bacteria do not form resting structures. They live in or on the host tissues in living or decaying rate.

4. **Survival through seeds**: Seed can harbor a wide range of microflora, viruses and other causal agents of plant diseases. The longevity of seed borne pathogens can be independent of the seeds they inhibits and depends on the capability of the pathogen to remain viable and infective from one season to next in or on their seeds. Pathogen may live even longer than the seeds that they colonize. For instance, *Colletotrichum lindemuthianum* remained active even after the bean seeds it colonized lost their viability. Flax seeds retain germianbility for 18 to 24 months, whereas *Botrytis cinerea, Colletotrichum linicola* and *Aureobasidium lini* survived for more than four years in seeds. The tobacco ring spot virus remained viable in soyabean seeds for five years at 16 to 32°C even most seeds failed to germinate.
5. **Survival through insects**: Survival of plant pathogens in association with insects is not uncommon in the nature. Some vectors such as aphids and mites carry non persistent viruses externally on or near the mouth parts. Viruses are known to persist between crops within vectors, which are themselves inactive while in case of persistent viruses the vectors remain infective much longer, some times throughout life or infectivity is passed to next generation. This provides an opportunity to virus to manage continuity of infection cycle. Barley yellow mosaic virus can persist up to ten years in resting spores of its vector (*Polymyxa graminis*) and tobacco rattle virus in dormant nematodes for about a year. The bacterium associated with corn wilt (*E. stewartii*) is present in the intestinal tract of its vector. Nematodes play vital role in survival and dispersal of many important plant viruses.
6. **Survival through nematodes**: At least twenty plant viruses have been found perennating with soil inhabiting, ectoparasitic nematodes. Nematodes of the genera *Longidorus* and *Xiphinema* acquire the viral pathgoens of tobacco ring spot, tomato ring spot, wheat spindle streak mosaic etc, diseases while those of genera *Trichodorus* and *Paratrichodorus* acquire viral pathogens of diseases like tobacco rattle and pea early browning.
7. **Survival through fungi**: Some viral pathogens have been found surviving in association with certain fungi. *Olpidium*, is the source of survival and transmission of viral pathogens causing tobacco mosaic, cucumber necrosis, lettuce big vein, and tobacco stunt diseases. Viruses causing wheat mosaic and beet necrotic yellow vein diseases survive in a fungus, namely *Polymyxa* while those causing disease potato mop top survive in *Spongosproa.* These viruses are borne internally in or externally on the resting spores and the zoosproes of their respective fungi. Such resting spores and zoosproes introduce viruses in host plants when they infect the latter.

19

Types of Parasitism and Variability in Plant Pathogens

A. Types of parasitism

An organism that lives within or upon another living organism from which it derives nourishment and in which it may cause various degrees of injury is called as *parasite*. The removal of food by a parasite from its host is called parasitism. Among the parasite one can distinguish different degree of parasitism as

i) Obligate parasites

ii) Non obligate parasites

iii) Facultative saprophytes

iv) Facultative parasites

i) Obligate Parasites or Biotrophs

Those organisms which requires living hosts or tissues to complete their life cycle , are called as obligate parasites or An obligate parasite is an organism that can live only on living tissue They can never be grown on dead, artificial food material. e.g. Rust, Mildews, Viruses.

ii) Non Obligate Parasites or Necrotrophs

Those organisms when kills the tissue in advance of penetration and then lives on it as saprophytically. e.g. *Sclerotium rolfsii, Ventruria, Claviceps.*

iii) Facultative Saprophyte

An organism that is ordinarily parasitic but under proper conditions may be saprophytic. e.g. *Smut, Sphacelotheca* sp.

iv) Facultative Parasites

An organism that is ordinarily saprophytic but under proper conditions may be parasitic. e.g. *Pythium, Phytophthora.*

B. Variability in plant pathogens

Variability in different organisms is common feature and is an important process for the continuity of life under stress conditions. Plant pathogens especially fungi, bacteria, virus and nematode show variations in their pathogenic potential (virulence) and physiological functions that support their survival and perpetuation in different environmental conditions. These variations may be largely due to genetic factors that may be conditioned by environmental factors. Various studies on genetic variability of microbial plant pathogens have shown that variations may occur in characters which affect their ability to infect host plants and also other characters that do not alter the pathogenic potential, but they may be concerned with their survival.

Knowledge of variation in pathogenic characteristics is valuable in planning effective measures for disease management. In addition, development of resistant to fungicides has been a serious concern to growers, researchers as well as the administrators, since the decision to continue application or withdrawal of a fungicide can affect both the growers and industry.

Mechanism of Variability

It is the property of an organism to change its characters from one generation to the other.

Mechanisms of Variability

S.No.	Fungi	Bacteria	Virus
1.	Mutation	Conjugation	Attenuation
2.	Recombination	Transformation	Mutation
3.	Hybridization	Transduction	Recombination
4.	Parasexuality		Selection Pressure
5.	Heterokaryosis		
6.	Hetreoploidy		
7.	Adaptation		
8.	Saltation		

Mechanisms of Variability in Fungi

1. **Mutation:** Mutation is one of the major causes of variability in the absence of sexual process. It is a more or less abrupt change in the genetic material of an organism. The frequency of mutation differs from biotype to biotype. Mutation factors are mostly recessive. Mutations for virulence occur in plant pathogenic fungi. Such mutants have been reported in *Phytophthora infestans, Puccinia graminis* and other fungi. Fungi of class ascomycetes are most suitable for mutation studies.

There are mainly two mechanisms of variability

i) General machanisms of variability: Two mechanisms of variability, namely, mulation and recombination, occur in pathogens.

ii) Specialized mechanisms of variability: Specific mechanisms for generating variability appear to operate only in certain kinds or organism. There specialiazed mechanism of variability include heterokaryosis, heterploidy and parasexualism in fungi, conjugation, transformation and transduction in bacteria, and genetic recombination in viruses.

2. **Recombination:**Most changes in the characteristics of pathogens are the result of recombination occurs during sexual processes. Recombination of genetic factor occurs during meiotic division of zygote as a result of cross over in which part of chromatid of one chromosome of a pair are expressed with that of the other. Recombination can also occur during mitotic division of cell in the course of growth of the individual and is important for creating variation in fungi.

3. **Parasexuality :** Parasexuality in fungi can be called *genetic recombination without meiosis*. In the absence of meiosis during the life cycle of imperfect fungi, recombination of hereditary properties and genetic variation still occur by a mechanism called parasexuality. Parasexuality has been reported in many rusts, including *P.graminis tirtici, P. coronata* and in some smuts *Ustilago hordei* and *U. maydis*. In rust fungi as *P. graminis tritici*, mitotic recombination may represent a most important method of generating new races especially in countries such as India where sexual stage of the fungus is rare due to scarcity of the alternate host, the barberry.

4. **Heterokaryosis:** The phenomenon of existence of different kinds of nuclei in the same individual is known as heterokaryosis. It has been demonstrated in numerous Ascomycetes, Basidiomycetes and Fungi Imperfecti. Hansen and Smith (1932) first suggested heterokaryosis as a means of variation in *Botrytis cinerea*. The pathogenicity is influenced by heterokaryosis, has been demonstrated in *Fusarium oxysporum f. pisi*, the incitant of pea wilt. Heterokaryosis is known to occur in wheat rust fungus.

5. **Heteroploidy:** Heteroploidy is the existence of cells tissues or whole organisms with numbers of chromosomes per nucleus that are different from the normal 1N or 2N complement for the particular organism. Heteroploids may be haploids, diploid, triploid, tetraploids or aneuploids, i.e. have one or more extra chromosomes or are missing one or more chromosomes from the normal euploid number e.g. N+1. Heteroploidy has been repeatedly observed in fungi and has been shown to affect the growth rate, spore size and rate of spore production , hyphal color, enzyme activities, and pathogenicity.

6. **Adaptation:** When a pathogen acquires ability to carry out a physiological process that it could not carry out before or could not carry out efficiently, it is said to have adapted and this phenomenon is known as adaptation. Adaptations in pathogenicity have been reported in *Fusarium*, *Phytophthora* and *Helminthosporium* spp.
7. **Satation:** Saltation is expressed by morphologically distinct areas or sectors in a fungal colony. These sectors may be remain stable or saltate again. Saltants have been reported in *Fusarium, Helminthosporium, Alternaria, Phoma* and *Sphacelotheca* spp. Mechanism of saltation is not clearly understood.

Mechanism of variability in Bacteria

The genetic recombination processes lead to the variability in bacterial population, are:

1. Conjugation
2. Transformation
3. Transduction

1. **Conjugation:** Genetic material of one cell is transferred to another cell during conjugation. The two cells are genetically different. The donor cell transfers part of its genome to the recipient cell. Conjugation is one type of genetic recombination. Genetic recombination refers to any process leading to the formation of a new individual which derives some of its gene from one parent and some from another, genetically different parent. Conjugation increases the genetic variability of bacterial populations and facilitates the emergences of antibiotic resistance.
2. **Transformation:** DNA from one type of bacterium is incorporated into the genetic makeup of another organism during transformation. In this DNA is absorbed through external source, This is a method employed in laboratory to bring about variation through recombination. It may be occurring in nature as well.
3. **Transduction:** Virus that infects bacteria is called *bacteriophage*. The infected bacterial cell bursts and the bacteriophage particles are liberated. Some of the bacteriophage particles released carry some genetic material of the host bacterial cell. They enter into another bacterial cell and the genetic material brought into the cell by the virus is incorporated into that of the bacteria. By this process a new genotype of the bacterium arises and this process is called transduction.

The transfer of genetic information in this manner is not always limited to members of the same species or even genus (vertical in heritance) for instance, gram negative bacteria can readily transmit genetic material across species, Agrobacterium transmits genes across kingdom barrier to plants. Such events are called horizontal gene transfers.

Mechanism of Variability in viruses

Variability in virus is due to

1. Recombination
2. Mutation
3. Selection pressure
4. Attenuation

1. **Hybridization/Recombination:** Plant viruses have always been regarded to undergo recombination. When two strains of the same virus are inoculated into the same host plants, one or more new virus strains are recovered with properties (symptomatology, virulence and so on) different from those of either of the original strains combinants. The new strains probably are recombinations. For example. two tobravirus strains called N_5 and 16 are formed by natural recombination of $RNAs_2$ of tobacco rattle and pea early browning viruses. Formation of new strains also occurs in brome mosaic virus by genetic recombination.

2. **Mutation:** Mutation is also an important source of origin of mutants. Mutations can be base substitutions and deletion mutations. Base substitution results in point mutation whereby base of a nucleotide is changed so that encoded protein molecule of parent strain. Such base substitutions have been discovered in cauliflower mosaic viruses and tobacco mosaic. Deletion mutations occur when part of the genome is deleted so that the mutant has lost the character encoded by the deleted segment of genome. Deletion mutations were first reported by Reddy and Black (1974, 1977) in wound tumor virus. Since the deletion mutations have been discovered in several other viruses.

3. **Selection Pressure:** Selection pressure resulting in the selection of new strains under changed conditions. With a change in selection pressure , an earlier minor strain of the mixture becomes the new dominant strain which seemingly leads to origin of new strains. Temperature and host are the two most common and important selection pressure operative in nature leading to the selection of new strains.

4. **Attenuation:** Attenuation means partial or complete loss of virulence in viruses. Attenuation also increases the genetic variability of viral populations and facilitates the emergences of new strain/biotypes.

Nematodes

Individual produced as a result of sexual reproduction are expected to be different from each other and from their parents in a number of characteristics which can be expressed morphologically or physiologically. The morphological characters like head shape, and scelerotisation, type of oesophagus, type of stylet and stylet knobs, position of vulva, number and types of ovaries, and neuroreceptors like amphids and phasmids, number of lateral lines, body dimensions and tail shape are important for the taxonomic identification of a specimen. The physiological variations may include preference of a host by a nematode species found in particular environmental and edaphic conditions where as the other members of the same species may not be host specific to that corp.

Stages of variation in pathogens

1. **Species:** The enitre population of a particular pathogen, for example a fungal pathogen, has certain morphological and phenotypic characteristics in common and make the speiecs of pathogen. Example: *Puccinia graminis*
2. **Varieties/Special forms:** Some individuals of this species however attack only wheat or barley or oats and these individuals make up groups that callled varieties or special forms (formae specialis). Example *P. graminis f.sp. tritici, P. graminis hordei etc.*
3. **Race:** Evan within each varieties/formae specialis, some individuals attack some of the varieties of the host plant but not the others, some attack another set of host plant varieties and soon with each group of such individuals making super. Example : There are more than 200 races in *Puccinia graminis tritici* (race-1, race-15 and soon).
4. **Variant:** Occasionally, one of the off spring of a race can suddenly attack a new variety that was not infected before, this individual in called variant.
5. **Biotype:** The identical individuals produced asexually by the variant make up a biotype. Each races consists of one or several biotype (race 15A, 15B and soon).

Question

Short Answer

1. Name the mechanisms that help generating variability in pathogens.
2. In what respect the parasexual cycle differs from a true sexual cycle in fungi?
3. What is the basic difference between generalized transduction and specialized transduction?

20

Pathogenesis Role of Enzymes, Toxins and Growth Regulators

Pathogenesis: Role of enzymes

The chain of metabolic events that bring about the disease is pathogenesis. It involves the virulence of the pathogen, susceptibility of the plant and impact of the environment. It is called *disease triangle*. Penetration infection, invasion, colonization and reproduction are the events that occur during pathogenesis, and lead to development of disease.

The establishment of pathogen in the host and consequent development of disease involves a chemical interaction between the host and the pathogen. The pathogen employs offensive chemical weapons to break host barriers, which the host resists with all its might. Parasitism and disease resistance run parallel and are inseparable, A biochemical 'tug of war' ensures between the host and the pathogen, in which one tries to outwit the other. Enzyme, toxins and growth regulators are the important chemical weapons of the pathogen. Viruses, however, because of their basic nature, do not produce any of these chemicals.

Pathogens produce these substances either in normal course of their activities (constitutively) or when, they grow on certain substrates such as their host plants (Inducible). Plant pathogenic enzymes disintegrate the structural components of host cells, breakdown inert food materials in the cell, or affect components of the protoplast directly and interfare with its functioning system. Toxins seem to act directly on protoplast component and interfare with the permeability of its membrane and with its function. Growth regulators exert a hormonal effect on the cells and either increase or decrease their ability to divide and enlarge. Polysaccharides seem to play a role only in the vascular diseases, in which they passively interfere with the translocation of water in the plants or they may also be toxic.

1. Enzymes in Plant Disease

Enzymes are protein molecules that catalyze all the inter-related reaction in a living cell. Some enzymes are present in cells at all times (constitutive) while some are produced only when needed (induced).

Role of important enzymes in pathogenesis

a) Cutinase

- Many fungi and a few bacteria have been known to produce cutinases.
- The cutinase enzyme produce by pathogens degrades cutin layer.
- The avirulent isolate did not produce cutinase.
- The amount of cutinase released by germinating spores of Fusarium solani f.sp.pisi, the pea rot was correlated with the degree of virulence of various isolates.
- All these observations suggest the importance of cutinase in disease development.

b) Pectinases

- Pectin methy lesterase (PME) and Pectin lyases polygalacturonases.
- These enzymes degrade middle lamella and cell wall and facilitates free movement of pathogens from cell to cell.
- Pectic enzymes release peroxidases from cell walls. *Peroxidase* induces browning of vascular tissues, a typical wilt syndrome in many plants.
- Pectic enzymes release endogenous *ethylene* from host cells. These ethylene causes loss of turgor, chlorosis and necrosis of leaves.
- Pectic enzymes causes vascular plugging producing gels. Swelling of gels leads to occlusion of vascular lumina. It leads to wilting of plants.
- Pectic enzymes in some cases cause cell death. *Erwinia carotovora* produces endo polygalacturonate trans –eliminase and this purified enzyme induces leakage from potato cell membrane and cell death.

Table. Cell wall and membrane degrading enzymes produced by certain plant pathogens

Enzymes	Substrate	Pathogen
Cutinases	Cutin	*Fusarium solani* *Colletotrichum gloeosporioides*
Cellulases	Cellulose	*Erwinia carotovora, Ralstonia, Aschochyta pisi, Rhizoctonia solani*
Hemicellulases	Hemicellulose	*S. rolfsii, E. carotovora, Penicillium expansum*
Proteinases	Protein	*R. solani, S. rolfsii,*
Ligninases	Lignin	*B. cinerea, Heterobasidium annosus*
Esterases	Suberin	*F. solani f. sp. pisi*

c) Cellulase

- Plant pathogens produces cellulase enzymes, which quickly degrade cell walls and facilitate the invasion of host cells by the pathogens.
- These enzymes become activated only after the action of pectic enzymes
- Cellulose degrading enzymes (cellulases) have been shown to produced by serveral phytopathogenic fungi, bacteria and nematodes.
- They facilitates the penetration and spread of pathogen in the host and causes collapse of cellular structure, thereby help in production of disease.

d) Hemicllulases

- Hemicellulases degrade hemicellulosic substances present in plant cells.
- The most important hemicellulosic enzymes are xylanase, glucanase, galactanase arabinare and mannare.
- *Sclerotium rolfsii* and Fusarim spp. are known to produce these enzymes.

However it is still not clear how they contribute to ability pathogen to cause disease.

e) Ligninases

- Ligninases are lignig degrading enzyme.
- White rot fungi secrete one or more enzymes (ligninases) which enable them to utilize lignin.

f) Proteases or Proteinases

- Invovles in degradation of cell wall structural proteins.
- Proteotytic enzymes secreted by pathogens can profoundally affect the organization and function of host cells.

g) Amylases

- The degradation of starch is brought about by the action of enzymes called amylases.
- The end product of starch breakdown is glucose which is used by pathogen directly.

h) Phospholipases/Lipases

- Lipolytic enzymes hydrolyze liberation of fatty acid from lipid.
- Only few pathogens produce these enzymes.
- Fatty acids are presumably utilized by the pathogen directly.

2. Toxins in Plant Diseases

Toxins

'Toxins' can be defined as low molecular weight, non enzymatic microbial products toxic to the higher plants. These are differs from other microbial products like 'mycotoxins' and 'antibiotics' which are toxic to the animals and microbes (except producers).

Classification

Toxins are classified as Phytotoxins, vivotoxins and pathotoxins.

Phytotoxins: Toxins produced by pathogens, which are toxic to plants but not considered as of primary importance during pathogenesis, is called as phytotoxin. Non specific toxins are considered as phytotoxins.

Vivotoxins: The toxin produced *in vivo* (in infected tissues) which function in disease development but not as the initial inciting agent, is termed as vivotoxin.

Pathotoxins: Pathotoxins are the toxins which are major determinants of disease and of pathogenicity. Pathotoxin induces all the typical symptoms in reasonable concentration and is co-related with pathogenicity. Host specific toxin can be considered as pathotoxins.

In recent classification, toxins are divided into two categories.

1. **Host non specific toxin**: A toxin which may affect many unrelated plant species in addition to main host of the pathogen producing toxin; it includes phytotoxin and vivotoxin.
2. **Host specific toxin**: A toxin which affects only the specific host of the pathogen; it includes pathotoxins.

Toxins in general, interact with cell membrane or organelles (chloroplast or mitochondria) and alter their permeability. Major host specific and host non specific toxins are listed in Table.

Table. Host Specific and non host specific toxins

Toxin	Pathogen	Host	Chemical nature
Host Specific toxins			
AK-toxin	*Alternaria kikuchiana*	Pear	Altenin (furanose ring)
HV- toxin	*Helminthosporium victorae*	Oats	Victoxinine
HC-toxin	*Helminthosporium carbonum*	Corn	Cyclic peptide
AM- toxin	*Alternaria mali*	Apple	Alternariolide
AB-toxin	*Alternaria brassicae*	Brassica spp.	Destruxin B
PC-toxin	*Periconia circinata*	Sorghum	Proteinaceous
Host Non specific toxins			
Tentoxin	*Alternaria alternata*		Cyclic tetrapeptide
Tabtoxin	*Pseudomonas tabaci*	Tobacco	Aminoacid derivative
Alternaric acid	*Alternaria solani*	Potato	Hemiquinone derivative
Cerato-ulmin	*Ceratocystis ulmi*		Large M. carbohydrate
Pyricularin	*Pyricularia oryzae*		Nitrogen containing compound
Fusaric acid	*Fusarium spp.*		5-n-butypicolinic acid

Primary determinant: *The toxin required for pathogenicity* is called *primary determinant.* It is essential for colonization by the producing pathogens and disease development in the host. The primary determinant is otherwise called *pathogenicity factor.*

Secondary determinant: *Thoe toxin required for virulence* is called *secondary determinant.* Secondary determinant accounts for certain symptoms or contribute to virulence, but are not essential for colonization. The secondary determinant is otherwise called *virulence factor.*

Pathogenesis: Role of toxins

- Toxin act directly on living host protoplasts, seriouslly damaging or killing the cells of the plant.
- Some toxins act as general protoplasmic poisons and affect many species of plant representing different families.
- Other are toxic to only a few plant species or varieties and completely harmess to others.
- Fungi and bacteria may produce toxins in infected plants and in culture medium.
- Toxins are extremelly poisonous substances and are effective in very low concentrations.
- Certain toxins act as antimetabolites and induce a deficiency for an essential growth factor.

- Toxins ensure host cells eighter by affecting the permeability of cell membranc or by inactivating or inhibiting enzymes and subsequently interrutting the corresponding enzymatic reactions.

Virulence is a quantitative term denoting relative disease inducing ability of a microorganism. Pathogenicity refers to the capacity of an organism to induce disease. A change in host range indicates a change in pathogenicity while a change in severity of disease indicates a change in virulence. But a question arises, when does pathogenesis begin and how much must the pathogen spread for colonization to be successful. Toxin is not required for penetration. Both toxin produced and no produced penetrate the host tissues. Initial colonization (upto24h) of both the isolates is also same. Only subsequent invasion is different. It is most likely that toxins play important role only as virulence factor rather than pathogenicity factor. In other words toxin is important in necrotrophic phase rather than in biotrophic phase. All the toxins isolated so far are from saprophytic pathogens. No toxin has been isolated from obligate parasites like powdery mildews, downy mildews and rusts.

3. Growth regulators in plant disease

Plant growth is regulated by a small numbers of groups of naturally occurring compounds that act as *hormones* and are generally called *growth regulators*. The most important *growth regulators* are *auxins*, *gibberellins*, and *cytokinins*, but other compounds, such as *ethylene* and *growth inhibitors*. Alternation in concentration of growth regulators have been found to be associated with several plant diseases particularly in those resulting in abnormal plant/organ/ cell growth like, galls, tumors, knot, stunting, curling, hypertrophy, hyperplasia, etc. Increase in level of a growth regulator may be either owing to its induced production (by pathogen or host) or inhibited degradation. Decrease in concentration is due to the enhanced degradation by the host's or pathogen's enzymes. Biotrophs like powdery mildews employ these hormones to draw their nutrients from the host cells.

Pathogenesis: Role of growth regulators

- Pathogens often cause an imbalance in the hormonal system of the plant and bring about abnormal growth responce.
- Pathogen can cause disease through secretion of growth regulators in the infected plant or through their effects on the growth regulatory system of the infected plant is made evident by the variety of abnormal growth response such as stunting, overgrowths, rosetting, excessive root branching, malformation, leaf epinasty etc.

Alternations in the level of growth regulators seems to be a consequence of pathogenesis. However, at least in three diseases hormones have been demonstrated to be determinant of pathogenicity. These are crown gall caused by *Agrobacterium tumefaciens,* (cytokinin and auxin), oleander and olive knot caused by *Pseudomonas syringae sp. savastanoi* (auxin), fasciation disease caused by *Corynebacterium fasciens* (cytokinin). In all these three bacteria genes have been identified and cloned which are responsible for the hormonal production. Deletion of these genes result in conversion of virulent strains into avirulent.

Table. Certain Plant diseases involving altered concentration of growth regulators

Growth Regulators	Disease (Pathogens)
Auxins	**Increased concentration**: Crown gall (*A. tumefaciens*), White stem rust (*Puccinina graminis tritici*), Downey mildew of crucifers (*Peronospora parasitica*), Tomato wilt (*Verticillium albo-atrum*), Banana wilt (*F. oxysporum f. sp. cubense*), Bacterial wilt of solanaceous crops (*Pseudomonas solanacearum*)
	Reduced concentration: Potato leaf roll(PLRV), Curly top of sugarbeet (SBCTV), Tobacco mosaic (TMV)
Gibberellins	**Increased concentration:** Bakanae disease of rice (*Gibberella fujikuri*), Creeping thistle rust *(Puccinia punctiformis*)
Cytokinins	**Increased concentrations**: Club root of crucifers (*Plasmodiophora brassicae*), Crown gall (*Agrobacterium tumefaciens*), White rust of crucifers(*Albugo candida*), Root knot of tobacco (*Meloidogyne incognita*)
	Reduced concentrations: Root knot of tomato (*M. incognita*), Verticillium wilt of tomato and cotton
Ethylene	**Increased concentration:** Tomato wilt (*F. oxysporum f.sp. lycopersici*), Verticillium wilt of cotton (*V. albo-atrum*), Wilt of tulip (*F. oxysporum f. sp. tulipae).*

Questions

Short Answer

1. Name two cutinolytic and two cell wall degrading enzymes secreted by the pathogens.
2. What is tissue maceration? What does it contribute in disease development.
3. Why sapwood but not heartwood of angiosperm and wood of gymnosperm is usually degraded by white rot fungi?
4. Give four principles on which non-host-selective (NST) activity relies upon.
5. Who first confirmed tabtoxin production in which plant disease? what was the casual organism of that disease?

6. Give the chemical composition and structure of victorin (HV-toxin).
7. Which fungal disease is related with the discovery of gibberellins? What was the pathogen of the disease?
8. What may be the usual causes for little or no water translocation in vascular wilt diseases?
9. Who reported the first case of altered permeability in diseased tissue? Name the disease and its pathogen.
10. Name some enzymes that are thought responsible for membrane damage in diseased plants.
11. What do you understand by 'extra haustorial membrane (EHM)'?
12. Name two disease with their causal agents in which the photosynthetic area is usually very much reduced.
13. Name two theories proposed to explain the increase in respiration rate in diseased plants.
14. Write about the role of enzymes in pathogenesis.

Long Answer

15. Write notes on:
a) Pathotoxins and vivotoxins.
b) Lignolytic enzymes in disease development.
c) Criteria of degradative enzymes involvement in pathogenesis.
d) HC-toxins.
16. Write in brief about the chemical weapons used by pathogens in pathogenesis.
17. Write an essay on the role of toxins in disease development.
18. Discuss the mechanism of respiratory increase in diseased plants.
19. Write notes on:
a) Physical blockage of xylem vessels.
b) Exceptions to the general pattern of reduction in photosynthetic activity in diseased plants.
c) Protein metabolism in diseased plant tissue.
d) Effect on transcription by pathogens.
e) Give a concise account of growth physiology in diseased plants.

21

Defense Mechanism in Plants

Introduction

Resistance to disease in a host plants is a condition in which the plant conquer the pathogen's attack and thus suffers small or no injury. Resistant hosts prevent or slow the development and reproduction of the majority of pathogen propagules that they come into contact with. Different plants defend themselves in different ways. Each kind of plant, probably, employs different defense mechanism against each of the pathogens that attack it. The defense barriers erected by plants are a co-ordinated system of molecular, cellular and tissue-based responses to pathogen attack.

There are various mechanisms of defense used by plants. For convenience, such mechanisms can be considered under two broads heads;

1. **Pre existing or Passive defense (Defense present on or inside the plant since beginning)**

a) Pre existing structural defense

b) Pre existing biochemical defense

2. **Post infection or Active defense (Defense in response to attack by the pathogen)**

a) Post infection structural defense

b) Post infection biochemical defense

1. Pre existing or Passive defense

a) **Pre existing structural defenses:** The external or internal structural barriers existing before pathogen attack are called *pre* existing defense structures or *passive* or *static* or *anti infection* structures. Initiating from periphery, the following barriers occurring in different host plants are described.

i) **Wax and cuticle:** The cuticle covers the epidermal cells of plants and consists of cutinized layer, pectin layer and a wax layer. The cuticular wax reduces the chances of infection by many pathogens.

ii) **Epidermal layer**: Thick and tough wall of epidermal cells may directly prevent the entry of pathogen completely or make it difficult. The toughness is due to polymers of cellulose, hemicelluloses, lignin, mineral substances, suberin etc. For instance, most outer wall of epidermal cells of rice plants are *lignified* and are seldom penetrated by *Magnaporthe grisea* that causes blast diseases of rice while the walls of *motor cells* are *proteinaceous* rather than lignified and the pathogens enter the host mainly through them.

iii) **Natural openings**: Natural opening such as *stomata*, *lenticels, hydathodes, nectarines* etc. allow many plant pathogenic fungi and bacteria to enter plants through them, but in some cases the structures of these openings prevent pathogen –entry. For instance, Mandarin orange is resistant to *Xanthomonas axenopodis pv. citri* because of *broad cuticular lips* covering the stomata. Similarly, the size and internal structure of lenticels may play a defensive role against the pathogens. Hydathodes are the nectarthodes in inflorescence of many plants. They secrete sugary nectar and this serves as barrier to those organisms that cannot tolerate high osmotic pressure. *High hairiness* of leaves and pods in chickpea is resistant character against *Aschochyta rabei.*

iv) **Internal structures:** There are many pre existing structures inside the plant that prevent the entry of pathogen beyond them. For example, in wheat varieties containing sclerenchymatous tissues in relatively high proportion in stem possess certain degree of resistance to *Puccinina graminis f. sp. tritici*. *Pythium* spp. invade only juvenile tissues , seedlings become resistant as they advance in age and secondary thickenings are formed.

b) Pre existing biochemical defense

Plants possess certain metabolic processes in their cells which imparts resistance to them against pathogen attacks. This type of defense adopted by the plants is called "*biochemical defense*". When biochemical defense exist in the plant even before the attack of the pathogen. This is called *pre existing biochemical* defense.

I. Release of antimicrobial compounds

Plant commonly exude organic substances from above ground parts and roots, containing amino acids, sugars ,organic acids, enzymes, glycosides etc. Some of the compounds occurring in exudates are *inhibitory*. These inhibitory substances directly affect the micro organism or encourage certain group to dominant the environment and function as antagonists of the pathogen. For instance, certain varieties of linseed resist infection of *F. oxysporum f. sp. lini* because of the presence of *HCN* in their *root exudates*. Presence of *HCN* in *roots* of sorghum and maize is also reported. *HCN* in sorghum root exudates might be responsible for reduction of *pigeonpea wilt* caused by *Fusarium udum* in plots where sorghum is grown as mixed crop with pigeonpea. **Red scale** of onion (red varieties) conatin *protocatechuic acid* and *catechol*, which may exude in drops and impart resistance to attack of *Colletotrichum circinanas* (onion smudge disease).The root exudates favor development of plant growth promoting rhizobacteria (PGPR), which suppresses the plant pathogens and induces resistance.

II. Inhibitor present in plant cells

The plant cells contain certain pre-existing inhibitory substances which mainly play a defensive role against the particular pathogen. *Phenols* and *quinones* are two classes of antimicrobial compounds produced by some plants Lactones, cyanogenic glucosides, saponins, terpenoids, stilbenes and tannins are also plant-produced compounds associated with pathogen resistance. Another group of defensive compounds are the plant *defensins*, which interfere with pathogen nutrition and retard their development. Secreted defensins can create an antimicrobial microenvironment for germinating seeds and accumulated defensins can provide defence against insect-transmitted viruses in flowers, leaves and tubers.

III. Lack of essential factors

a) **Recognition factors**:Plant may not infected by a pathogen if there surface cells lack specific recognition factors. If the pathogen does not recognize the plant as one of its hosts it may not adhere to the host surface or it may not produce infection substances such as enzymes or structures like appresoria and haustoria. These recognition molecules are of various types of oligosaccharides, polysaccharides and glycoproteins.

b) **Essential nutrients and growth factors**: The host specialized pathogens (obligate parasites) particularly among the fungi need certain specific substances and these substances are only available in adequate quantities in the host they infect. Hence, absence of these specific substances makes the other varieties inappropriate host for such a pathogen. Example, A fungus, namely, *Venturia inaequalis* causes apple scab disease. Certain races of this pathogen recovered have lost their ability to synthesize a growth factor in them due to mutation. Races so mutated prove to be non pathogenic to apple.

c) **Common antigen in host plant**: Presence of common antigen (protein) in both host and pathogen determines disease occurrence in the host. But, if, the antigen is absent in the host and present only in the pathogen, it makes the host resistant to that particular pathogen. For instance, the varieties of linseed which have an antigen common to their pathogen *Melampsora lini* prove to be susceptible to rust disease. In contrary, the linseed varieties not having the antigen in them but occurring in the pathogen are resistant to the pathogen. Similar is the case with leaf spot disease of cotton caused by *Xanthomonas compestris pv. malvacearum.*

2. Post infection or Active or Induced defense

a) Post infection or Induced structural defense

After the pathogen successfully penetrates the pre-existing defense barriers of the host, the latter, some how, generally develops definite defense structures inside them in response to the reaction taking place between host and pathogen. These structures so developed are generally of four types:

- Induced histological defense
- Induced cellular defense
- Induced cytoplasmic defense
- Induced hypersensitive defense

I. **Induced histological defense :**Certain defense structures are developed inside the host as a result of tissue differentiation or chemical substance deposition in tissues ahead of or around the pathogen. These defense structures are called induced histological defense. These may be at various levels:

II. **Cork layer formation**: Cork layer formation is a part of natural healing system of plants. Cork layers blocks the flow of nutrients and water from the healthy to the infected area and this deprives the pathogen of nourishment resulting in its weakness or, sometimes, death. Further, it also block inward flow of toxic substances produced by the pathogen so that such toxic substances could not damage the underlying healthy tissues of the host. Example. Common scab of potato (*Streptomyces scabis*) and *Rhizophus* rot of sweet potato.

III. **Tyloses formation**: The tyloses are formed by protrusion of xylem parenchymatous cell walls, through pits into xylem vessels. The formation of tyloses can also restrict the spread of pathogenic propagules in the xylem, although they also tend to reduce the movement of water through the vessels, causing water stress in the plant. Examples: in formation of tyloses sweet potato against *Fusarium oxysporum f. sp. batatas.*

IV. **Gum deposition**: Wounded tree trunks often secrete protective gums that seal the wound from further infection. Example: blast disease of rice (*Magnaporthe grisea*), Leaf spot of rice (*Helminthosporium oryzae).*

V. **Abscission layers:** It is the gap between host cell layers and devices for dropping-off older leaves and mature fruits. Plant may use this for defense mechanism also, i.e. to drop-off infected or invaded plant tissues or parts, along with pathogen. Shot holes in leaves of fruit trees is a common characteristic.

VI. **Induced cellular defense:** The cellular defense structures, i.e. changes in cell walls, have only a restricted role in defense. Following types are generally observed:

- Carbohydrate apposition (synthesis of secondary walls and papillae formation).
- Callose deposition (hyphal sheathing just outside plasma lemma around the haustorium, which delays contact of pathogen (*Phytophthora infestans*) with host cells.
- Structural proteins

b) **Induced cytoplasmic defense**: When the hyphae of a pathogen succeed in penetrating a particular cell, the cytoplasmic content of such cell becomes organized in such a way that the cytoplasm beomes granular and dense and certain organelle like structures develop in it. As a result, the penetrating hyphae disintegrate into small granular bodies and further development of the hyphae is stopped. Cytoplasmic defense structures are, perhaps, the very last line of structural defense and are effective against some slow growing pathogens, weak parasites or some symbiotic relationships.

Table.14. Post-infectional morphological and structural defense mechanisms against plant pathogens

S.No.	Defense barrier	Pathogen	Host
1	Formation of cork layer	*Rhizoctonia solani*	Potato
	Formation of tyloses	*Fusarium oxysporum f.sp. batatas*	Sweet potato
2	Formation of abscission layer	*Xanthomonas arboricola pv. pruni*	Sour cherry
3	Sheathing of hyphae	*Phytophthora infestans*	Potato
4	Gum deposition	*Pyricularia oryzae*	Rice

Hypersensitivity: Hypersensitive cell death is another widespread mechanism used by hosts to prevent the spread of a pathogen. Infected cells and those surrounding them “suicide”, preventing further spread, and in some cases, killing the pathogen.

b) Post infection or Induced biochemical defense

These mechanisms restrict the spread of the pathogen after infection is established and contain the damage to host tissues.

I. **Role of Phytoalexins:** *Phytoalexins are low molecular weight antibiotics produced by many (but not all) plants in response to infection.* Phytoalexins inhibit the growth of bacteria and fungi *in vitro* and *in vivo*, and production of these antibiotics during an infection can induce resistance to consequent infections by that pathogen. Over 350 phytoalexins are known in over 100 plant species. They include *pterocarpans*, *cryptophenols isocoumarins*, *isoflavenoids*, *sesquiterpenes*, and others. Phytoalexins may be produced by any part of the plant, although different phytoalexins can accumulate in different organs. Normally, related plant species produce structurally-related phytoalexins, and many produce more than one, enabling the plant to present a toxic cocktail to invading pathogens. Phytoalexins are produced in cells surrounding an infection site and delivered to the infected cell packaged in lipid vesicles, creating a toxic micro-environment in the infected cell and, hopefully, preventing disease establishment. Phytoalexin accumulation is often associated with *hypersensitive cell death*, although only living cells can synthesise phytoalexins. Some plants can also sequester phytoalexins into vacuoles as stores of inactive sugar-conjugates, which can be cleaved and released quickly if initial defence responses are unsuccessful.

Table 15. Some important phytoalexins

S.No.	*Phytoalexins*	*Producer Plant*	*Chemical nature*
1	*Resveratrol*	*Vitis vinifera*	*Stilbene*
2	*Trifolirhazin*	*Trifolium pratens*	*Isoflavonoid*
3	*Isocoumarin*	*Carrot root*	*Isocoumarin*
4	*Rishitin*	*Potato tuber*	*Sequiterpene*
5	*Pisatin*	*Pisum sativum*	*Isofalvonoid*
6	*Maackiain*	*Cicer arietinum*	*Isofalvonoid*
7	*Phaseollidin*	*Phaseolus vulgaris*	*Isofalvonoid*

II. **Role of phenol compounds**

Phenolics are the common compounds found in plants and possess an aromatic ring bearing a hydroxyl sustituent. Phenolics are toxic to fungi, bacteria and viruses. They inhibit:

- Mycelia growth
- Spore germination

- Producion of cellulotic and pectic enzymes
- Activity of cellulotic and pectic enzymes and
- Production of toxins by the fungi

Some of the individual phenolic substances have been identified as the key defense chemicals in plants. They are:

Phenolic substances	Crop
Chlorogenic acid	Potato
Catechol and Pyrocatechic acid	Onion
Avenacin	Oats
Hordatine	Barley
Tomatine	Tomato
Tuliposides	Tulip
Catechin and Isoquercitin	Cotton
Umbelliferone and Scopoletin	Sweet potato

III. **Role of Pathogenesis-related proteins (PRPs)**

Pathogenesis-related proteins are defined *as plant proteins that are induced in pathological and relate situations.* Their presence before infection increases resistance to pathogens. These novel proteins are accumulated maximally 7-10 days after infection, and indicate the attainment of the systemic acquired resistance by the plant. The PRPs accumulate in the intercellular spaces and in the vacuole. The intercellular PRPs present first line of defense to a challenging pathogen, and if this fails and tissue is disrupted , the release of vacuolar PRPs act as second line of defense, engulfing pathogen with lytic enzymes, which have *antifungal*, *antibacterial* and *antiviral* activity.

Table.16: Families of pathogenesis –related proteins (PRPs)

S.No.	Protein family	Protein activity	Targeted pathogen sites
1	PRP-1	Unknown	Active against omycetes
2	PRP-2	β-1,3 glucanase	Cell wall glucan of fungi
3	PRP-3	Endochitinase	Cell wall chitin of fungi
4	PRP-4	Endochitinase	Cell wall chitin of fungi
5	PRP-5	Taumartin-like	Active against omycetes
6	PRP-6	Proteinase inhibitor	Active on nematodes and insects
7	PRP-7	Endoroteinase	Microbial cell wall dissolution
8	PRP-8	Endochitinase with lysozyme activity	Cell wall chitin of fungi, and mucopeptide wall of bacteria

9	PRP-9	Peroxidase	
10	PRP-10	Ribonuclease activity	Viral RNA
11	PRP-11	Endochitinase	Cell wall chitin of fungi
12	PRP-12	Defensin	Antifungal and antibacterial activity
13	PRP-13	Thionin	Antifungal and antibacterial activity
14	PRP-14	Lipid transfer proteins	Antifungal and antibacterial activity
15	PRP-15	Oxalate-oxidase	Produces H_2O_2 that inhibits microbes and also stimulates host defense
16	PRP-16	Oxalate-oxidase-like with superdismutase	Produces H_2O_2 that inhibits microbes and also stimulates host defense

IV. **Inactivation of enzymes and toxins:** The role played by enzymes and toxins of pathogens during pathogenesis is well established. The *hemibiotrophs* and *necrotrophs* employ more of these substances for causing host tissue damage as compared to biotrophs. The defense strategy of resistant plants, through activity of phenols, tannins, proteins as enzymes inhibitors. the phenolics are not anti–fungal but make pathogen ineffective by neutralizing their enzymes.

Table 17. Defense through detoxification of pathogen toxin

Detoxifying compound	Pathogen toxin	Disease	Pathogen
Chlorogenic acid	Piricularin	Blast of rice	*Pyricularia oryzae*
Dehydroascorbic acid	AK toxin	Black spot of pear	*Alternaria kikuchiana*
High concentration of cysteineFusaric acid		Wilt of cotton	*Fusarium oxysporum f. sp. vasinfectum*
		Wilt of tomato	*Fusarium oxysporum f. sp. lycopersici*

Systemic Acquired Resistance (SAR) & Induced Systemic Resistance (ISR)

Features	ISR	SAR
Definition	Resistance activated in response to artificial application of non pathogenic strains of pathogen, PGPR, and abiotic stresses	Resistance activated in response pathogenic infection.
Pathway	Jasmonic acid dependent pathway	Salicyclic acid dependent pathway
Signalling molecule	Jasmonic acid	Salicyclic acid
Defense gene involved	CHIB, HEL, PDF1,2	PR-1 and PR-2
Pathogen associated	Operates against biotrophic pathogens	Operate against necrotrophic and facultative parasite

Questions

Short Answer

1. What do you understand by 'constitutive' mechanisms of defence?
2. Szinkum variety of citrus is resistant to citrus canker causing pathogen. Which defence strategy helps do this?
3. Through which passage in lingnified epidermis the blast disease causing pathogen manages to enter inside rice plants?
4. Why the white-scale varieties of onion are susceptible to smudge disease but the red-scale varieties are resistant to the same?
5. What do you understand by 'oxidative burst'?
6. Who gave the basic concept of phytoalexins? Which were the first phytoalexins identified?
7. What is systemic acquired resistance (SAR)? Which virus is credited to have been involved in SAR-discovery?
8. Write about role of toxins in pathogenesis.

Long Answer

9. Give a summarized account of the post–infection biochemical defence mechanisms in plants.
10. Attempt the following:

a) Hypersensitive response

b) Tyloses and vascular wilt diseases

c) Phytoalexin-biosynthesis in plants

d) Pathogenesis-related proteins

e) Induced systemic resistance (ISR)

11. Discuss difference type of structural defence strategies developed in response to pathogen-attack in plants.

12. Explain the following

a) Gene Deployment

b) Pyramiding of genes

c) What is plant defense? White down the various mechanisms of defense used by plants

d) Describe the different phases of infection process

22

Epidemiology Factors Affecting Disease Development

Epidemiology

Vanderplank has called epidemiology the *science of disease in population*. Epidemiology puts its emphasis on *population phenomena.* The important populations are those of the host, on the one hand, and of the pathogen, on the other. But further, epidemiology is concerned with *host-pathogen interactions* that lead to disease and crop losses. These interactions are often environmentally dependent. So epidemiology deals with effects of the biotic and abiotic environments. Finally , man has had important effects on hosts, pathogens and the environments, effect that today are increasing in magnitude and scale of importance. Epidemiology therefore also deals with the effects of man on disease and sometimes with the effects of plant disease on mankind. All of this can be symbolized by the *disease tetrahedron.*

Purpose of study of epidemiology

The study of epidemiology serves at least two purposes;

- Understanding of the behavior of disease in time and space.
- Using that understanding to manage disease.

Factors affecting disease developemnt

The following table hightight the major factors affecting disease development.

Pathogen	Host	Environment
Presence of pathogen	Type of crop	Temperature
Mode of dispersal of pathogen	Degree of genetic uniformity of host plants	Rainfall / Dew
High birth rate of pathogen	Genetic make-up	Leaf wetness period
Low death rate of pathogen	Host age	Soil properties
Survival efficiency	Host nutrition	Wind
Reproductive fitness	Host population structure	Fire history
Adaptibility of pathogen	Nature of propagation	

1. Host factors

a) **Type of crop**: Epidemics generally develop much more rapidly in *annual crops* (cereal. pulses, oilseeds and vegetable crops) than *perennial crops* (fruit and forest trees). Some epidemics of fruit and forest trees, for instance, pear decline citrus tristeza, dutch elm and chestnut blight disease, takes year to develop.

b) **Degree of genetic uniformity of host plants**: A severe plant disease epidemic may develop when genetically uniform plants are grown over large area, greater likely hood exists that a new pathogen race will appear that can attack their genome and result in an epidemic.This phenomenon has been observed in repeatedly, for instance, in southern corn leaf blight on corn carrying Texas male sterile cytoplasm. The *lowest* rates of epidemic development generally occurs in *cross pollinated* crops, intermediate rates in self pollinated crops and the *highest rates* in *vegetatively* propagated crops.

c) **Level of genetic resistance or susceptibility of the host**: Vertical and horizontal resistance have distinct bearing on development and progress of epidemic. Hosts plants carrying horizontal resistance will probably become infected, but the rate at which the disease and the epidemic will develop depends on the level of resistance and the environmental conditions, Host plants carrying *vertical resistance* do not allow a pathogen to become established in them, and thus *no epidemic* can develop. Susceptible host plants lacking genes for resistance against the pathogen provide the ideal substrate for establishment and development of new infections.

d) **Host age**: Plant change in their susceptibility to disease with *age*. For instance, damping off and roots rots, downy mildews, peach leaf curl, rusts, smuts, bacterial blights and viral infections, the host plants are susceptible only during their growth period and becomes resistant during adult stage. In some other diseases such as flower or fruit blights caused by fungal pathogens like *Alternaria, Penicillium, Glomerella* and *Monilinia* and in all post harvest infections, plant parts (fruits) are resistant during growth and early adult stage but become susceptible near riepning.

e) **Host nutrition**: Balanced nutrition is required for proper growth and development of plants. Under good nutritional conditions, biotrophic fungi have a relative advantage; under poor nutritional conditions necrotrophic and perthotrophic fungi have a relative advantage. The type and amount of fertilizers have influence on the development of plant diseases.

f) **Host population structure:** Crop rotation, sequence and crop density have major bearing on the type and amount of plant diseases. High crop density will influence microclimate and will make infection and dispersal of inoculums more effective. Both these factors will help the development of disease.

g) **Nature of propagation**: Nature of propagation affects the development of disease. Chances of disease are more if plants are vegetatively propagated.

2. Role of pathogen factors

There are certain important pathogen factors that affect the course of epidemic

a) **Level of virulence:** Virulent pathogens capable of rapidly infecting the host ensure faster production of larger amounts of inoculums, and thereby, disease, than pathogens of lesser virulence.

b) **Mode of dispersal of pathogen**: Rapid and easy dispersal of the pathogen inoculums from its source to healthy hosts in the field may help frequent and wide spread epidemic. Epidemics of diseases like mildew, rusts and leaf spots occur usually as the spores of their pathogen are released into the air and get disseminated quickly by it over distances up to several miles. Contrary to it, soil borne pathogens like *Fusarium, Verticilium*, etc., and most nematodes usually have 2 to 4 reproductive cycles per growing season. Since, the dispersal of such pathogens is limited both in space and time, only localized and slower developing epidemics are caused.

c) **High birth rate of pathogen:** High birth rate or fast reproductive cycle of pathogen is key factor for establishment & development of epidemics. Some fungal pathogen have capacity to produce enormous quantity of spores asexually and these spores, if dispersed fast over long distances and favored by conducive weather conditions, can result in epidemics in their respective areas.

d) **Low death rate of pathogen**: Epidemics attributed to low death rate of pathogens are those in which the causal organism is systemic and protected by the plant tissues. Thus, the chances of high mortality are considerably reduced and ultimately greater chance of widespread infections in a given area exist.

e) **Adaptibility of pathogen:** This is another factor vital to development of epidemics. For pathogens having capacity to acclimatize to adverse conditions, the occurrence of epidemics is about certain. The units of propagules produced by the pathogen are dispersed by external agencies which must be available if epidemics are to develop.

E. Role of Environmental factors

The presence of a pathogen against a particular plant will generally not cause serious disease unless the environmental conditions are favourable. This includes the aerial environment and the soil (*edaphic*) environment.

i) Aerial environment

Properties of the aerial environment that influence disease development include, temperature, moisture levels and pollution.

a) **Temperature**: Temperature affects the *incubation*, or latent, *period* (the time between infection and the appearance of disease symptoms), the *generation time* (the time between infection and sporulation), and the *infectious period* (the time during which the pathogen keeps producing propagules). The disease cycle speeds up at higher temperatures, resulting in faster development of epidemics. The period of leaf wetness, combined with temperature information can be used to predict outbreaks of some diseases (infection periods) and be used to time preventative treatments, such as spraying.

b) **Moisture:** Moisture is mainly important to pathogenic fungi and bacteria. Rain splash plays significant role in the dispersal of some fungi and nearly all bacteria, and a period of leaf wetness is essential for the germination of most air borne spores. By using water for dispersal, propagules are dispersed at a time when they are likely to be able to germinate as well. Because the process of germination and infection takes time, the duration of leaf wetness also influences the success of the infection. The duration necessary for infection varies with temperature. Generally, a longer period of leaf wetness is desired to establish an infection in cooler temperatures, as germination and infection are normally accelerated in warmer conditions.

c) **Pollution:** A lately recognised aspect of the aerial environment that can influence disease in plants is air pollution. A high concentration of pollutants can affect disease development and, in extreme cases, damage the plants directly by causing acid rain.

ii) Soil environment

The soil environment affects soil-borne diseases, largely by determining the amount of moisture available to pathogens for germination, survival and motility. Germination and infection success also rely on the temperature of the soil. The fertility and organic matter content of the soil can affect the development of disease. Plant defences are weakened by nutrient deficiency, although some pathogens, such as rusts and powdery mildews, thrive on well-nourished plants. Other diseases thrive in soils that are specifically low in organic matter.

F. Role of humans activities

Humans/farmers have a direct and indirect influence on plant disease epidemics. Human decide the kind, the time, the number, and the density of plants cultivated in a particular area. If these aspects are not decided with precision, they may lead to the occurrence of an epidemic in that area. Similarly continuous monoculture, use of high dozes of fertilizers, excess of irrigation, and poor field sanitation are certain cultural practices adopted by man which invite severe epidemic in a given area. The planting of certain variety or the use of certain chemical may lead to selection of virulent strains that either can attack the resistance of the variety or are resistant to the chemical and thus lead to epidemics.

Types of epidemics

Monocyclic epidemics: Monocyclic epidemics are caused by pathogens with a *low birth rate* and *death rate* meaning they only have one infection cycle per season. They are typical of soil born diseases such as *Fusarium* wilt of pigeonpea.

Polycyclic epidemics: Polycyclic epidemics are caused by pathogens capable of *several* infection cycles a season. These are most often caused by airborne diseases such as powdery mildew.

Polyetic epidemics: Some diseases takes *several years* from the time of infection until symptom develop and pathogen reproduction. Epidemics that occur under these conditions are referred to as *polyetic* epidemics and can be caused by both *monocylcic* and *polycyclic* pathogens. Apple powdery mildew is an example of a polyetic epidemic caused by a polycyclic pathogen and Dutch Elm disease a polyetic epidemic caused by a monocyclic pathogen.

Differences between monocyclic and polycyclic plant disease epidemics

Characters	Monocyclic	Polycyclic
Birth rate	Low	High
Death rate	Low	High
Rate of increases	Like simple interest	Like compound interest
Survival	Long lived	Short lived
Reproduction	Reproduces only one generation per growing season	Reproduce new batches of propagules several times per growing season

Epidemiologists use specific terminology to accurately communicate magnitudes and potentials for disease.

- **Disease progress curve**: Periodic observations on disease severity , in any field or area, when plotted against time on a graph paper gives curved lines representing progress of disease over time . These cumulative disease curves are known as Disease progress curve (DPC).
- **A saturation curve**-Monocyclic disease.

- **A sigmoid curve**- Polycyclic disease.
- **A bimodal curve**-Polycyclic disease affecting different organs.
- **Disease-gradient curve;** A graph that plots disease vs. distance from an inoculum source.

Questions

Long Answer

1. What is epidemiology? Do epidemiology of plant disease is considered within the scope of agricultural meteorology.
2. How epidemics of plant disease are caused? Describe the various factors leading to the development of epidemics of late blight of potato and wheat rusts.
3. What are slow and rapid epiphytotics ? Give the environment related factors responsible for the establishement of an epiphytotic.
4. Give a summarized account of effect of environment on disease development.
5. Give in detail the effect of temperature and moisture on disease development.
6. Write notes on:

a) Effect of host, nutrition on disease development.

b) Effect of light on disease development.

c) Effect of O_2 and CO_2 concentration on disease development.

Short Answer

1. Name four diseases (with their pathogen) which are influenced by soil temperature.
2. Why humid weather accompanied with low temperature helps spread of late blight of potato in severe form?
3. Name some diseases that are favoured by alkaline soil.
4. How deficiency of nitrogen favours disease development?
5. Name some plant pathogens that cause infection at low relative humidity.
6. Why root rot disease of cotton occurs in sandy or loamy soil while wilt of cotton occurs in heavy clay soils?
7. Name four diseases which occur in alkaline soil.

23

Forecasting, Prognosis, Modeling and Expert System in Plant Pathology

Disease Forecasting

Definition

Forecasting involves all the activities in ascertaining and notifying the grower of community that conditions are sufficiently favourable for certain diseases, that application of control measures will result in economic gain or on the other hand, and just as important that the amount expected is unlikely to be enough to justify the expenditure of time, energy and money for control. The above statement made explicit distinction between *positive forecast* and *negative forecast* and both have value for growers as well as society in general.

Positive forecast: Employs need based chemical sprays, provides adequate protection to crop and reduces damage to environment.

Negative forecast: Avoids unnecessary chemical sprays, no risk to the crop health and no disruption of environment.

Importance of forecasting

- Forecasting informs the growers that the conditions are not favourable and the disease is unlikely to be intense enough. This helps growers to save the expenditure in terms of energy, time and money by not applying unnecessary management practices.
- It informs the growers about when the conditions are going to be sufficiently favourable for economically important diseases of an agroclimatic zone. This help growers to apply disease management timely that results in economic gain.
- It helps growers to plan advance preventive measures against likely losses due to the occurrence of economically important severe diseases.
- Government and other organizations initiate necessary steps for the timely stocking of chemicals, equipments, etc. when they are timely warned by the possibility of an epidemic by the forecasting's.

Basic requirements for disease forecasting

The following basic requirements are essential to make useful and successful disease forecasts

- Disease must cause economically significant damage in terms of yield loss or quality.
- Disease should not be regular feature.
- Reliable means of communication with farmers.
- Control measures must be available at an economically acceptable cost.
- Growers must have sufficient manpower and equipments to apply control measures when disease warning is given.
- Farmers should be adaptive and have purchase power.

Information's needed for disease forecasting

The information's required for forecasting are:

1. Host Factors

a) Prevalence of susceptible varieties in the given locality

b) Response of host at different stages of the growth to the activity of pathogen

c) Density and distribution of the host in a given locality

2. Pathogen Factors

a) Amount of primary inoculum in the air, soil or planting material

b) Dispersal of inoculums

c) Spore germination

d) Infection

e) Incubation period

f) Sporulation on the infected host

g) Re-dispersal/Dissemination of spores

h) Perennating stages

i) Inoculum potential and density in the seed, soil and air

3. Environmental factors

a) Temperature

b) Humidity

c) Light intensity

d) Wind velocity

Methods of disease forecasting

Disease forecasting requires field observations on the pathogen characters, collection of weather data, variety of the crop and certain investigations and their correlations. Usually the following methods are employed in disease forecasting.

1. Forecasting based on amount of initial inoculum

Amount of primary inoculum present in seed, soil or vector can be estimated and disease prediction can be made without much problem. This criteria is more important for monocyclic diseases where the secondary infection does not take place and the amount of primary inoculums is related to disease severity and damage. It is being used in case of loose smut of wheat and soil borne inoculums of *Rhizoctonia, Sclerotium, Verticillium*, nematodes can be estimated and disease can be forecast. Presence of loose smut of wheat, ergot of pearl millet and viral diseases of potato can be detected in the seed lots at random by different seed testing methods. Seed testing methods can be used to determine potential disease incidence and enable decision to be made on the need for chemical seed treatment.

2. Forecasting based on weather conditions

Prevailing weather conditions *viz*., temperature, relative humidity, rainfall, light, wind velocity etc., have become a major criteria for diseases forecasting for polycyclic diseases where in addition to amount of primary inoculums, the multiplication and dispersal of secondary inoculums are weather dependent. Forecasting of late and early blight of potato, wheat rust, anthracnose, apple scab, powdery mildew of cucurbits and downy mildew of grapes are made considering the prevailing weather conditions which determine the amount of disease and damage of the crop.

3. Forecasting based on correlative information

Weather data of numerous years are collected and correlated with the intensity of the diseases. The data are compared and then the forecasting of the disease is done. Forecasting criteria developed from comparisons of disease observation with standard meteorological data have been provided for diseases like fire blight of apple and powdery mildew of barley.

4. Use of computer for disease forecasting

The forecasting of disease is also made by the use of computers. The computers are fed with weather data collected on the farm by individual farmers. The computer then processes the data, determines whether an infection period is imminent, likely to occur or can not occur, and a recommendation is made to the farmer as to whether or not to spray and what materials to apply. This system gives the results quickly. One such computer based programmes in the USA is known as 'Blitecast' for potato late blight.

Examples

a) Late blight of potato

Late blight of potato had been instrumental in development of diverse fundamental concepts in plant pathology and disease forecasting is no exception. The potato growers in Europe were intelligent enough to recognize blight weather (Moderate temperature and lots of moisture: rain, dew or humidity) much earlier than *Van Everdingen* postulated *Dutch rules* in 1926. Followings are four conditions correlated with weather that could predict about appearance of late blight in Netherlands:

- Night temperature below dew point at least 4 hours.
- Minimum temperature of 10^0C or slightly above.
- Clouds on the next day.
- Rainfall during next 24 hours of at least 0.1 mm.

Notable contributions in late blight forecasting

- Van Everdingen postulated Dutch rules in 1926
- Krause *et al.* (1970), BLITECAST
- Singh *et al.* (1980), JHULSACAST for Indo Gangetic plains
- Fry and Apple (1983), Integrated host resistance and fungicide weathering in BLITECAST
- Dommermuth (1988), Phytoprog I. late blight warning service
- Runno and Kopple (2002), NEGFRY
- Grunwald (2002), Modified and validated SIMCAST

Prognosis

Prognosis is characterized as the *prediction of the outbreak, development* and *outcome* of disease. The objective is to decide in advance whether expected damage is threatening and whether control measures are to be taken. Prognosis has been attempted at two levels:

a) Date prognosis

b) Loss prognosis

a) **Date prognosis**: mainly deals with the *prediction of the out break of a disease*. Following two criteria are considered for date prognosis:

- Estimate the initial population of pathogens and their antagonist, if feasible.
- Record weather parameters and their prediction over prognosis time span.

b) **Loss prognosis:** is based on *estimate of expected economic loss*, in relation to disease intensity-crop damage, and constitutes the basis or the decision as to spray or not. In loss prognosis in addition to economic aspects, consideration of the biological and eco toxicological consequences of a plant protection measure are important.

Modeling in Plant Pathology started with pioneering synthesis of Vander plank (1963) and aerobiology of Gregory (1961). *Models* are *simple representation of otherwise more complex real systems*. The model provide opportunity for better *understanding* and *system management*. Kranz (1974) stated that *a model* might be *verbal statement*, a *hypothesis*, a *theory* or a *law*. The selection of suitable model is determined by the purpose and goals which must be defined very precisely.

1. **Conceptual Models:** These models were conceived during early stages of introduction of models in Plant Pathology. They were fundamentally qualitative, without much quantification. The relationships and interactions were displayed in diagrammatic form or flowchart manner. examples, Disease pyramid, Components of pathogenesis, gene for gene hypothesis.

2. **Analytical Models:** Vanderplank (1963) used differential equations for calculation of pathogen population growth. Equation of single and multiple cycle diseases, threshold theorem and fitting curves to empirical data are common analytical models used in plant pathology.

3. **Predictive Models:** are those in which the emphasis is placed on those variables which have the greatest value in forecasting the likely course of events. These are generally used in *estimation of yields* and *forecasting diseases*. Regression and differential equations are generally employed. They are considered a boon to growers for bringing precision in management decisions. Examples, BLITECAST in USA and PHYTOPROG in West Germany in giving shot term warnings of potato late blight attacks. FAST is the computer model for forecasting early blight in tomato.

4. **Simulation Models:** The simulation models try to *simulate* or mimic the real life situations under study. When a computer is provided data describing the diverse subcomponents of the epidemic and measures of control at specific points in time, it then provides back the continuous information regarding not only the spread and severity of the disease over time, but also the final crop and economic losses likely to occur due to disease under the conditions of the epidemic as provided to the computer. The first computer simulation programme, called EPIDEM, was written in 1969 to simulate early blight epidemics of tomato and potato. Examples of other simulators are:

Name of the programme	Disease	Pathogen
EPIPRE	Cereal rusts and aphids	*Puccinica graminis tritici*
EPICORN	Southern corn leaf blight of maize.	*Helminthosporium maydis*
CERCOS	Leaf blight of celery	*Cercospora spp.*
MYCOS	Blight of chrysanthemum	*Mycosphaerella spp.*
EPIVEN	Apple scab	*Venturia inaequalis*
EPIDEMIC	Stripe rust	*Puccinia striiformis*
TOM-CAST	Early blight of tomato	*Alternaria solani*
BLIGHT-CAST	Late blight of potato	*Phytophthora infestans*
PLASMO	Downey mildew of grapes	*Plasmopara viticola*
SIMCAST	Late blight of potato	*Phytophthora infestans*
NEGFRY	Late blight of potato	*Phytophthora infestans*
EPIVET	Viral diseases of potato	Contact and aphid transmitted viruses

Expert Systems

An agricultural *Expert System* is a *Decision Support System* for Agricultural Extension Agents who have to decide what advice to be offered to farmers who in turn decide about suitable action based on it. It is one of the most efficient extension tools to *take the technology from scientists* to *the farmers directly without any dilution of content* which normally creeps in because of the number of agencies involved in normal technology transfer systems.

Expert systems are the frontiers, which combines and integrate both *science* of *plant pathology* & *art of diagnosis* and *disease management*. These computer decision support systems were formally introduced in Plant Pathology in 1987. Expert system are computer programs that emulate the *logic* and *problem* solving proficiency of human expert. The computerized disease forecasting systems of 1970s (BLITECAST) and 1980s (apple scab forecaster) were the precursors to expert system. In principle the Expert system are an *artificial intelligence* application that uses encoded knowledge from a human expert. Expert system has the ability to use complete or incomplete data; accordingly it will assign certainty values to the solutions of the problems. The Expert system are programmed to review a consultation and provide user with an explanation. Three main areas involved in decision making are *pest risk estimation, current threat assessment* and *pesticide application.*

Expert System used in Plant Pathology

Expert System	Description
PLANT/ds	First expert system (1983) developed for diagnosis of 17 soybean diseases (USA)
EPIPRE	Epidemic prediction and prevention for spring and winter wheat
POTATOES	Late and Early blight of potato
TOMEX	Diagnosis of 37 diseases of tomato
POMME	For management of disease and insects of apple
PSACO	For management of diseases and insects of apple
PRO-PLANT	For cereals, potato, sugarbeet and vegetables
More Crop	For management of rusts and other problems
TOMEx –UFV	For diagnosis of tomato diseases
SIMPHYT-III	For late blight of potato

24

Principles of Plant Disease Control

The techniques used in disease management mainly based on two principles: *Prevention or prophylaxis* and *cure.* The first principle prevention includes *disease management tactics applied before infection* i.e., *the plant is protected from disease.* It includes: avoidance, exclusion, eradication, resistant & protection. Second principle curative action: *functions with any measure applied after the plant is infected* i.e., *the plant is treated for the disease.* It includes *therapy.*

Principles of plant disease control

Prophylaxis	Avoidance	Prevent disease by selecting a time of the year or a site where there is no inoculum or where the environment is not favorable for infection
	Exclusion	Prevent the introduction of inoculums
	Eradication	Eliminate, destroy or inactivate the inoculums
	Protection	Prevent infection by means of a toxicant or some other barrier to infection
	Resistance	Utilize cultivars that are resistant to or tolerant of infection
Curative	Therapy	Cure plants that are already infected

1. **Avoidance of pathogen** : *Avoiding disease by planting at times when or where inoculum is absent or ineffective due to unfavorable environment conditions.*

- Choice of geographic area
- Selection of field
- Choice of time of sowing
- Disease escaping varieties
- Selection of seed and planting stock
- Modification of cultural practices

2. **Exclusion of inoculum**: *Preventing the inoculum from entering or establishing in the field or area where it does not exist.*

- Seed treatment
- Inspection and certification
- Quarantine
- Eradication of insect vectors

3. **Eradication of pathogens**: *Reducing, inactivating, eliminating or destroying inoculum at the source, either from a region or from an individual plant in which it is already established.*

- Biological control of plant pathogen
- Crop rotation
- Removal or destruction of diseased plant organs
- Rouging
- Eradication of alternate and collateral hosts
- Sanitation
- Heat and chemical treatment of disease plants
- Soil treatments

4. **Resistant:** *Utilize cultivars that are resistant to or tolerant of infection.* Resistant varieties, If available you can manage disease at zero cost
5. **Protection measures**: *Preventing infection by creating a chemical toxic barrier between the plant and the pathogen.*

- Chemical treatment
- Chemical control of insect vectors
- Modification of environments

6. **Therapy**: *Reducing severity of disease in an infected individual.*

- *Chemotherapy:* is the application of chemicals to an infected or diseased plant that stops the infection.Examples: heat or chemical treatment of vegetative material such as bulbs, corms, and woody cuttings to eliminate fungi, bacteria, nematodes or viruses that are established within the plant material.
- *Heat therapy*: Hot water treatment of rice seed at51-53°C for 15 min after dipping for 1 d in cool water for management of white tip disaesce caused by *Aphlenchoides besseyi*.
- *Tree surgery*: Use of chaubatia paste after pruning of diseasesd parts in horticultural crops.

While these principles are valid today, in the context of modern concepts of plant disease management, they have some critical shortcomings like

- The principles of plant disease control are stated in absolute terms like *exclude, prevent* and *eliminate* that imply a goal of *zero disease*. Plant disease control in this sense in not practical, and in most cases is not even possible. Indeed, we need not eliminate a disease; we merely need to reduce progress and keep *disease development below an acceptable level.* Instead of plant disease control, we need to think in terms of *plant disease management.*
- These principles donot take consideration the *dynamics of plant disease*, that is the changes in the incidence and severity of disease in time and space. We need some means of assessing *quantitatively* the effects of various control measures, singly and in combination on the progress of disease.
- Finally, these principles of plant disease control tend to emphasize *tactics* without fitting them into an *adequate overall strategy.*

Does this mean that we should abandon the traditional principles? Of course not; We merely have to fit them into an appropriate overall strategy based on epidemiological principles.

Strategies versus Tactics

The dictionary definitions of the two terms are similar, but generally speaking, *an overall plan for reaching a particular objectives is called strategy*, while the *specific means for implementing a given strategy* are called *tactics.* What is a strategy at one level of focus could be called a tactic at another level of focus

Strategy of disease management

It is well accepted that complete eradication of a pathogen from the earth of the ecosystem is very difficult and against the law of nature. Therefore, the strategy for disease management should be to reduce *the population of pathogen below permissible level* so they do not cause damage. Keeping this in mind, the strategy of disease management should integrate all known methods taking into account the following factors.

1. Pathogen inhibition
2. Tolerable losses
3. Economical control
4. Long term control
5. Collective approach

1. Pathogen inhibition

As complete eradication of a pathogen is not possible, there should be more emphasis on to develop a technology which *maintain their population below the damaging level.*

2. Tolerable losses

It means tolerate the level of disease that do not harm economically. For example, in crops raised for seed two per cent disease infection is permissible, whereas in crops raised for consumption the permissible limit is 5 per cent.

3. Economical control

The strategy for disease management should be implemented after considering the *cost benefit ratio*. The return per unit of money spent must be sufficiently high.

4. Long term control

The strategy of disease management should attempt to utilize methods which have a *long term effect* so as to reduce the expenses and pollution of the environment.

5. Collective approach

Disease control cannot be achieved by a single farmer since pathogen can move from field to field. Hence, *collective area wise approach is necessary* for effective control.

Any war cannot be won only on quantity of weapons available. Success only comes from a well planned strategy involving the best use of weapons in a coordinated manner at the right time and with proper methods. Similarly, the war against diseases can be won provided an *integrated approach is employed and all the five basic factors discussed above are taken into account.*

Questions

1. Distinguish between i. Control and Management ii. Prevention and Cure iii. Strategies vs Tactics.
2. Why the term management is preferred over the term control?
3. Enlist principles of plant disease management.
4. Write down the major critical shortcoming of principles of plant disease management.

5. What is therapy. Name some fruit trees which can be controlled by tree therapy.
6. Discuss in details the general principles of plant disease management with suitable examples.
7. Elaborate strategies of plant disease management .
8. What is plant disease management? Give the basis of the principles of plant disease management.
9. Distinguish between the following:

i) Exclusion and Eradication

ii) Protectant and Eradicant

iii) Prophylaxis and Immunization

iv) Chemotherapy and Thermothearapy.

25

Physical Management

Defintion

- *Management of plant diseases using physical means is called physical management.*
- Also called *physical therapy* or *thermotherapy.*
- It is mostly *curative* in nature.

Principles

The principles involved in physical management or thermotherapy is that the pathogens present in seed, soil, or planting material are.

- Inactivated or eliminated at temperatures nonlethal for the host tissues.
- Heat causes inactivation and non immobilization of the pathogen.
- The rate at which the pathogen is inactivated is determined by temperature, the higher the temperature, the faster is the inactivation.

Methods

Physical methods are employed for reduction or elimination of primary inoculums that may be present in seed, soil, or planting material are

i) Hot water treatment (HWT)

ii) Hot air treatment (HAT)

iii) Steam and aerated steam therapy (AST)

iv) Moist hot air treatment

v) Solar heat treatment (SHT)

i) Hot water treatment (HWT)

- Hot water treatment is widely used for the control of seed borne pathogens, especially bacteria and viruses.

- Main drawback in the hot water treatment is that the seeds may be killed or loose its germ inability, if the period of treatment exceeds the specified time.

Table 18. Control of seed borne pathogens through Hot –water treatment

Crop	Disease	Causal organism	Seed Treatment
Rice	White tip	*Aphelenchoides besseyi*	51-53°C for 15 min after dipping for 1 d in cool water.
Pearl millet	Downy mildew	*Sclerospora graminicola*	55⁰C for 10 min
Cauliflower & Cabbage	Black rot	*Xanthomonas campestris pv. campestris*	50⁰C for 20 min

ii) Hot air treatment (HAT)

- Hot air treatment is less effective than hot water treatment but less injurious to seed and easy to operate.
- It has been used against several diseases of sugarcane and is employed for treating canes which are soft and succulent.
- Hot air treatment at 54⁰C for 8h, effectively eliminates RSD pathogen without impairing the germination of buds.
- Similarly, grassy shoot disease of sugarcane has also been controlled by hot air at 54⁰C for 8h.

iii) Steam and aerated steam therapy (AST)

- The use of aerated steam is more effective than hot air and safer than hot water in controlling seed borne infections.
- The heat capacity of water vapour is about *half that of water* and 2.5 *that of air*, hence air temperature and time required may be higher than that of hot water and lower than that of hot air.
- The advantages of this method include easier drying of seeds, low loss in germination, easy temperature control and no damage to seed coat of legumes.
- Sugarcane setts are also exposed to aerated steam at 50⁰C for 3 hrs to eliminate mosaic virus.

iv) Moist hot air treatment (MHAT)

- This method is effectively used in sugarcane to eliminate grassy shoot disease.
- Initially the setts are exposed to hot air at 54⁰C for 8 hrs, then exposed to aerated steam at 50°C for 1 hr and finally to moist hot air at 54⁰C for 2 hours.

v) Solar heat treatment (SHT)

- Solar heat treatment is effective in controlling both *seed borne* and *soil borne* diseases
- It has been devised in India to eliminate the pathogen of *loose smut* of wheat.
- Luthra in 1953 devised a method in India to eliminate the deep seated infection of *ustilago nuda.* The method is popularly known as *solar heat or solar energy treatment.*
- In this method the seeds are soaked in cold water for 4 hours in the forenoon on a bright summer day followed by spreading and drying the seeds in hot sun for four hours in the afternoon. Then, the seeds are again treated with carboxin or carbendazin at 2g/kg and stored.
- This method is highly useful for treating large quantities of the seed lots.

vi) Soil Solarization

- Defined as *hydrothermal treatment of soil* for management of soil borne diseases.
- In this management tactics, the solar energy is preserved with the help of transparent white polythene sheet for about 40 days to increase soil temperature (10-15^{0}C) above normal temperature enough to kill the most of the soil borne diseases.

Results of solarization

- Increased soil temperature.
- Improved soil physical and chemical features.
- Control of disease and pests.
- Encouragement of beneficial soil organisms.
- Increased plant growth.
- Fungal diseases such as damping off, root and stem rots, wilt and blights diseases and some nematodes have been successfully managed by soil solarization.

Refrigeration

- The low temperature at or slightly above the freezing point checks the growth and activities of all such pathogens that cause a variety of post harvest diseases of vegetables and fruits.

- Therefore most perishable fruits and vegetables should be transported and stored in refrigerated vehicles and stores to manage post harvest diseases.

Questions

1. What are the objectives of the physical methods of plant disease management? How are they helpful in minimizing disease incidence?
2. Who developed solar heat treatment device? Name the disease for which the device was developed.
3. Write short notes on i. Principle of physical plant disease control ii. Hot water treatment iii. Hot air treatment iv. Steam and aerated steam therapy.
4. Explain about moist hot air treatment and solar heat treatment.
5. Why aerated steam is more effective than hot air and safer than hot water in controlling seed borne infections?
6. What is soil solarization? Describe the method, advantage and disadvantage of soil solarization.
7. Describe the role of regrigeration and radiation in plant disease management.
8. What is physical control ? Write down the different physical methods used for plant disease control.
9. How can solar energy be used for the treatment of seeds? In which disease has it been used? Is it more conveineient than chemical treatment.
10. Distinguish between hot water treatment and moist heat air treatment.Give atleast four examples of plant disease controll by hot water treatment.
11. What is thermotheary? What is the scientific principle involved in thermotherapy.

26

Cultural Management

Introduction

Cultural management is defined as *manipulation /or adjustment of crop production techniques for management of plant diseases*. Cultural practices have been as old as possibly agriculture itself. In early stage of agriculture development, the growers through their experiences and observations had known that repeated cultivation of a particular crop species or variety on a piece of land often resulted in crop sickness. As a matter of fact, in the present day agriculture, cultural practices are being considered as essential backup methods for management of plant diseases. Adequate adjustment in crop production techniques can modify the environment in such a manner that it becomes unfavorable for the pathogen and pathogenesis. Based on this the disease control affected by cultural practices are *preventive*, these practices aim at reducing the activity and density of inoculum. It offer the *opportunity to alter the environment*, the *condition of the host, behavior of the causal agent to achieve economic management of disease.*

Procedures

The procedures for disease management through cultural practices are described under the following three heads:

A. Adjustment of cultural practices

B. Sanitation

C. Eradication

A. Adjustment of cultural practices

The main purpose of adjusting of cultural practices is to *seed healthy seeds in healthy soil and to obtain a healthy crop stand.* The procedure for disease management through adjustment of cultural practices are described under following heads

1. **Summer ploughing:** Deep ploughing during summer periods buries the inocula of fungi of soil borne nature. Fungal propagules, sclerotia and different

types of spores/conidia on plant refuses die when exposed to sunlight due to the higher temperature prevailing during the summer. Further, alternate hosts and alternative hosts, infected self-sown plants, volunteer host's plants, weed hosts and regrowths from the plant roots, are also destroyed. Here, the spread of the disease is avoided. Groundnut blight (*Corticium rolfsii*) is controlled by ploughing the soil to a depth of 20 cm.

2. **Adjustment of sowing time:** Properly adjusting the sowing dates can give good dividends. While choosing the time of sowing it should be taken into consideration that susceptible stage of the crop growth and soil conditions and other environments favourable for maximum activity of the pathogen does not fall at the same time. Late planted wheat crops attract less infection than wheat planted on normal dates. Avoiding cool and cloudy days for planting will help to reduce red rot of sugarcane. Rapeseed sown in mid to late august is more liable to attack by leaf spot (*Alternaria brassicae*) than late-sown crops. Pea and gram planted soon after rains when soil temperature and moisture are at a high level, show high incidence of root rot and blight. As the soil temperature falls and moisture becomes less (Nov-Dec) these diseases are also reduced. In areas where these diseases are serious, late sowing helps in saving the crop. Stem rust of wheat damages the late sown crop more than the early sown crop. Because, time of onset of disease and ear formation coincides. Sowing from January to April or October to December is advocated to escape from the attack of neck blast of finger millet. Peas and chickpea sown in october usually suffer heavily from root rot and wilt (a complex of *Fusarium, Rhizoctonia* and *Sclerotium*). When these crops are sown late, the diseases are not so severe or almost absent. The groundnut rosette is transmitted by *Aphis craccivora*

3. **Selection of seeds and seed materials:** Seeds and seed materials form the primary source of infection. The absence of an initial inoculum in seeds is definitely helpful in delaying or suppressing the incidence of the disease. It is a preventive method. The diseases like foot rot, brown spot, short smut of sorghum, loose smut of wheat, bacterial blight of rice, bacterial blight of cotton, leaf crinkle of blackgram etc., are transmitted through seeds. Virus diseases and black ring of potatoes, foot rot of ginger, foot rot of betelvine, panama disease of banana, red rot of sugarcane, cassava mosaic, bunchy top and virus diseases of fruit trees are transmitted through tubers, setts, rhizomes, corns, grafts and budwoods. *Tuber indexing* 'is a special method to obtain disease free seed materials in potato. It is commonly practiced by nurseries and seed merchants selling potato seed tubers. Use of seeds in the place of rhizome/sucker is recommended in the control of katte' disease of cardamom.

4. **Provision of drainage facilities:** Water stagnation in different patches of field favours the fungi like *Pythium, Phytophthora*, *Rhizoctonia solani*, etc., for which proper leveling of the field before sowing or planting is very essential. Further improving the drainage is necessary in the control of sheath blight of rice. Provision of drainage channels in orchard crops like citrus, jackfruit, mango etc., in the garden is necessary before planting. In the control of damping-off diseases of vegetable and other crops, raising seedling in the raised beds method is followed. Foot rot of ginger (*Pythium myriotylum*) is also controlled by following the raised bed system of nursery.

5. **Depth of sowing:** Depth of sowing greatly influences seed transmitted diseases. Shallow planting in wet soils protects wheat plants from *Urocystis tritici* (flag smut) of wheat. Deep planting may cause delay in the emergence of seedlings, which may be vulnerable to pre-emergence damping off. Early emergence results in early lignification of tissues which become resistant to attack of soilborne pathogens.

6. **Spacing:** Closer spacing invariably alters the microclimate underneath the canopy of the crop which may provide favourable environment for development of diseases. Boll rot in cotton is quite common in crowded crop. Closer spacing favours many air borne diseases because of high humidity in the crop canopy. Early and late blight of groundnut and blister blight of tea are more in dense canopy. Early spread of black rot of cabbage takes place in closer spacing. Crowded stands may reduce some systemic diseases. Cotton wilt caused by *Verticillium alboatrum* will be less in closely planted crop if the fungal inoculum is less in the soil. Similarly closer spacing of rice reduces rice tungro virus infection particularly when vector population is less. Avoiding shade and providing wider spacing reduces the incidence of powdery mildew of tobacco. Late blight of potato and downy mildew of grapevine spread fast in closer spaced crops. The virus of tomato leaf curl, transmitted by *Bemisia tabaci*, is less severe in a crowded planting than in spaced planting.

7. **Altering the soil pH :** The altered pH of the environment forms a barrier against the pathogen. A very low pH less than 5.2 is unfavourable to common scab bacterium on potato.On the other hand the club root pathogen of cabbage cannot live and infect when the soil pH is 7.0 or more.

8. **Crop rotation:** Crop rotation is the oldest and cheapest method adopted in agriculture for eradication of certain types of pathogens from infested soil. Monoculturing provides opportunity for perpetuation of pathogenic organisms in the soil when the same crop is raised year after year in the same field. The soil-borne pathogens of that crop easily perennate in the soil and increase

in their population. After sometime, the soil becomes so heavily infested that it becomes unfit for cultivation of the particular crop. Virus diseases of crop plants and their vectors are found to increase after every crop if a crop is cultivated continuously in a field. On the other hand, when immune, resistant or non-host crops are grown for a definite duration after a susceptible crop in the field it is expected that in the absence of nutrition, the pathogen will be starved off and the population of such pathogens consequently decreases. Example, Crop rotation with sugarcane or paddy is effective in the control of panama wilt of banana (*Fusarium oxysporum* f. sp. *cubense*) and crop rotation with paddy or green manures is effective in the control of red rot of sugarcane (*Colletotrichum falcatum*). Rotation of cereal crops like pearl millet, finger millet or fox-tail millet is recommended for the control of Macrophomina root rot of pulse crops. Crop rotation is essentially a preventive measure and has its effect mainly on the succeeding crop.

9. **Mixed cropping:** Mixed cropping is growing of two or more crops simultaneously on the same piece of land. Mixed cropping materially helps in checking certain diseases. By proper selection of crops for the mixture, soil environment can be changed to one that is not favorable for the pathogen. Control of root rot of cotton by growing cotton with moth is an example. Further, due to reduced number of host pants in a mixed crop the susceptible area for an air borne foliar pathogen is decreased. Therefore, there is less primary infection and less production of secondary inoculums for spread of the disease. This slow down the rate of disease control. Blight of pulse crop (*Phyllosticta phaseolina*) has been successfully overcome by growing pulses as a mixed crop with cereals like sorghum and pearlmillet.

10. **Intercropping:** Intercropping is also a device in the control of some soil borne diseases. Intercrops should be properly chosen so that they should not have any common pathogen.Example, Intercropping of sorghum in pigeonpea field reduced the wilt (*F. udum*) incidence. The roots of non-host plants may act as a barrier obstructing the movement of pathogens in soil. They may release toxic substances from their roots which may suppress the growth of the pathogens attacking the main crop. *Hydrocyanic acid* (HCN) in root exudates of *sorghum* is toxic to *F. udum*, the pigeonpea wilt fungus.

11. **Barrier cropping:** Taller crops can be grown to protect a crop of lesser height from virus vectors. Barrier cropping with 3 rows of maize or sorghum or pearlmillet around the main crop namely blackgram or greengram is effective in reducing the vector population and incidence of yellow mosaic. Another best example is growing of 3 rows of kale or barley as barrier crops in cauliflower seed beds and undersown beet steckling against

cauliflower mosaic and beet yellows diseases, respectively. The incoming aphids are thought to land on the barley or kale and probe briefly, causing them to lose the non-persistently transmitted virus they are carrying. Maize or sunflower are the other barrier crops considered for these crops.

12. **Decoy crop and trap crop:** Decoy crops (hostile crops) are non-host crops sown with the purpose of making soilborne pathogens waste their infection potential. This is effected by activating dormant propagules of fungi, seeds of parasitic plants, etc. in absence of the host. Trap crops are host crops of the pathogen, sown to attract pathogens but destined to be harvested or destroyed before they complete their life cycle. Fodder sorghum can be raised as a trap crop to reduce downy mildew of sorghum.

13. **Mulching** :Mulching or covering of top soil with organic residues often helps in reducing plant diseases. Mulches of non-host origin should be used in the field. These mulches are known to release inhibitory substances in the underlying soil and also promote development of parasites and predators of nematodes. Reflective surfaces (mulches) laid on the soil around the crop plant, have been found to be highly effective in controlling aphid vectors. Aluminium strips or grey or white plastic sheets are used as mulch and it has successfully protected red peppers against CMV and PVY in Israel and summer squash against watermelon mosaic virus in the Imperial valley of California. Straw mulches have been successfully used to control the white fly – transmitted tomato yellow leaf curl virus in tomato crops in Israel. It is believed that the colour of the straw attracts the whiteflies and they are subsequently killed by the reflective heat. The disadvantage with straw mulches is that they eventually lose their yellow colour, but prolonged control may be obtained if straw is replaced by yellow polythene sheets.

14. **Irrigation water management:** Irrigation to the crop in the field is to wet the soil to the extent that roots easily get water and nutrients. If excess water is added to soil, it may directly affect activity of pathogens and/or it may affect disease incidence through the effect on the host. By irrigating the field, soil temperature is brought down, stress is removed and the disease is suppressed. When excess irrigation is made the juvenile stage of plants is lengthened making it susceptible to attack of fungi like Pythium. Supply of frequent but low quantity of irrigation water is, therefore, recommended for reducing chances of damping off in nurseries. Under conditions of excess water, respiration of roots is inhibited and many soluble salts accumulate in toxic amounts around the roots and base of the stem. This increases disease proneness of the roots. Irrigation increases guttation. Guttation drops on leaves serve as media for multiplication and penetration of many pathogens, such as *Helminthosporium* spp. on cereals and *Xanthomonas campestris*

on Brassica spp. Pathogens directly taking advantage of excess water are those that need wet soil for (i) activation of their resting structures and (ii) for movement of these propagules. Thus, in presence of excess free water bacterial cells and zoospores of pythiaceous fungi are dispersed easily. Therefore, at the plant stage when these pathogens can attack the crop irrigation should be avoided. Sclerotia, smut spores, chlamydospores, oospores and mycelium found in the soil are carried from one field to another through irrigation and drainage water. Stem rot, sheath blight and bacterial blight diseases of rice, damping off of vegetables and Macrophomina root rots of many crops spread mainly through irrigation and drainage water. Hence care should be taken not to irrigate a healthy field using drainage/irrigation water from a diseased field.

15. **Management of plant nutrients :**The plant nutrients in general when applied in excess may increase or reduce the resistance in plants to diseases. Increased application of nitrogenous fertilizers increases the incidence of many diseases. Crops fed with heavy doses of nitrogenous fertilizers grow robust with foliage and succulent tissue but become highly susceptible to the attack of diseases like rust powdery mildew, blast, tobacco mosaic and some bacterial diseases. Some diseases are favored by ammoniacal form of nitrogen while others are favoured by nitrate form of nitrogen. In general wilts (*Fusarium* sp.) and root rots (*Rhizoctonia* spp.) are favoured by *ammoniacal* nitrogen while verticillium wilts and root rots due to Pythium spp. are favoured by *nitrate* nitrogen. Repeated application of phosphatic fertilizers delays the onset and lessens the severity of take-all disease of barley (*Gaeumannomyces graminis*). Potassium application reduces the disease incidence in many crop diseases probably by increasing phenolics synthesis in plants. Application of potash induces resistance in groundnut against root rot caused by *Macrophomina phaseolina*. Calcium application suppresses the lesions due to the *R.solani* on bean roots. It is due to formation of calcium pectate, which is less available to action by polygalacturanase (PG) enzyme than is pectic acid. *Manganese* reduces late blight of potato, *ferric chloride* controls rice brown spot and *silicon* application reduced rice blast.

16. **Fallowing:** Fallowing starves the pathogen and helps in reduction of the inoculum by elimination of the host. Diseases like Macrophomina root rot on different crop plants is controlled by following this method. Wet fallowing makes the pathogenic propagule in or on the soil to germinate, spent them, is become susceptible attack of saprophytes. Example, *Sclerotium rolfsii* and *Verticillium dahliae*. The sclerotia or microsclerotia of these fungi are activated in the absence of root exudates of this host. They germinate quickly

when there is alternate wetting and drying of the soil. When the population of *Pythium myriotylum* is not high wet fallowing is successful in reducing the population. Wet fallowing reduces saprophytic survival of *Alternaria solani* on crop debris.

B. Field and plant sanitation

Field and plant sanitation is an important method of disease control through cultural practices. The inoculum present on field plants in the field may multiply on the plant or in the soil and in due course of time may be sufficient to nullify or reduce the effect of control practices. Many pathogens overwinter or oversummer on plant debris during the off-seasons and become active when the crop is again grown in the field. Hence plants bearing pathogens or plant debris introducing inoculum into the soil should be removed as early as possible. In most of the soil borne diseases like wilt and root rot, it has been reported that as long as the dead roots and other roots and other affected parts are present in the soil, the fungus continue its growth. When such diseased plant materials are removed, there is quick decline in the population of pathogens in the soil. In this manner Fusarium wilt of cotton, pigeonpea and banana, Verticillium wilt of cotton, root rot of beans, downy mildew of pearlmillet, sorghum, maize and peas, foot rot of betelvine, bacterial blight of cotton, white rust of crucifers, black spot of rose, powdery mildew of pea and cereals are reduced. In certain areas the linseed rust fungus (*Melampsora lini*), the rice blast and brown spot fungi and the fungus causing early blight of potato also perennate through dormant stages in diseased crop debris. Destruction of crop debris by burning immediately after harvest reduces the amount of inocula which survive through debris. Avoidance of the transfer of inoculum from one field to another by man, machine or water is one of the ground rules of cultural control. Where soil-borne diseases are concerned, anything that carries soil is suspect, this includes wheels, boots and water flowing either from adjacent fields, or through drainage ditches from distant fields. As regards sap-borne viruses, attention must be paid to disinfection of wheels and of the hands of labourers, as they pass from one field to another. Where such virus can also be carried on clothing. The work should be planned so that the labourers do not go from older to younger fields on the same day.

C. Eradication

Eradication is the elimination of pathogen after it has become established in the area where host is growing. The following are the important methods followed to prevent the spread of the disease:

i) Eradication of alternate hosts.

ii) Eradication of collateral and self-sown overwintering hosts.

iii) Eradication of the affected plants or trees.

iv) Eradication of pathogens from infected plant parts by surgery.

i) **Eradication of alternate hosts:** Removal of alternate hosts helps to prevent and check the spread of the disease caused by black rust pathogen in wheat. Barberry bush is the alternate host for black stem rust pathogen *Puccinia graminis tritici* on wheat where the pathogen survives in the offseason. The eradication of barberry had two benefits i.e. elimination of early spring primary inoculum and prevention of the formation of new physiologic races of the pathogens.

ii) **Eradication of collateral and self-sown overwintering hosts:** There are many weed hosts or wild species of cultivated plants act as *collateral hosts or volunteer plants* of an economic crop which act as reservoirs of pathogens of annual crop. Reservoir hosts help the pathogen to continue the infection chain. The primary inoculum is produced on and dispersed from these hosts to the cultivated crop hosts. If these wild or uneconomic host plants of the pathogen are destroyed, the sources of primary inoculum are eliminated and chances of initiation of the disease in the crop hosts are reduced. Destruction of these hosts breaks the life cycle of the pathogen and the infection chain.

iii) **Eradication of affected plants or trees:** It is essential to eradicate the host and the pathogen from an area in some threatening plant diseases. Citrus canker (*Xanthomonas axonopodis pv. citri*) is an example of success of an eradication programme. This disease was first noticed in Florida citrus trees in 1913. An eradication campaign was started in 1915. All the citrus nurseries and orchards were inspected and the infected trees were cut and burnt. The eradication programme continued till 1927 and no citrus canker was present in that area. Peach yellows and peach rosette were also controlled by removal and destruction of diseased trees.

iv) **Eradication of pathogens from infected plant parts by surgery:** Eradication of affected plant parts (tree surgery) are also practiced in certain cases which reduces the source of primary inoculum. Lesions caused by fire blight bacterium (*Erwinia amylovora*) on pear and apple trees are removed during winter months. This not only prevents further spread in the affected trees but also reduces the amount of inoculum that can spread to other branches and trees. Tree surgery is also practiced in coconut trees affected by stem bleeding disease (*Ceratocystis paradoxa*), citrus gummosis (*Phytophthora citrophthora*), *Dendrophthoe* spp. on citrus, bud rot of palms (*Phytophthora palmivora*) and koleroga of arecanut (*P. arecae*).

Questions

1. Discuss in details the plant disease management by modification of agricultural practices.
2. Which type of disease can effectively managmed by crop rotation?
3. What is the advantage of deep summer ploughing?
4. Describe the role of mixed cropping and fallowing in plant disease management.
5. What is cultural management? Write down the procedures for disease management through cultural practices.
6. Give a detail account on use of cultural practices for plant disease management.
7. Describes the role of field and plant sanitation in plant disease management.
8. Explain the following : i. Field and plant sanitation ii. Organic amendment of soil.
9. How plant diseases are influenced by macro and micro nutrients.
10. Name five plant diseases which can be controlled by field and plant sanitation.

27

Plant Quarantine

1. Biosecurity

Biosecurity is strategic and integrated approach encompasses policy frameworks to analyze and manage risks for food safety, animal life and health and plant life and health including environment.

Basic Elements

- Introduction of exotic pests/diseases.
- Emergence of virulent strains/races/ biotypes of economically important indigenous pests/diseases.
- Deliberate misuse of pests/pathogens.
- Biosafety issues arising from LMOs.

2. Plant Quarantine

The term 'Quarantine' is derived from the latin word *quarantum,* meaning 40. It refers to the 40 days period of detention of ships from countries with bubonic plague and cholera in the Middle Ages. The first such quarantine was imposed in Venice in 1374. *Plant quarantine , restrict entry of plants, plant products, soil, cultures of living organisms, packing materials, and commodities, as well as their containers and means of conveyance to protect agriculture and the environment from avoidable damage by hazardous organisms.*

3. Basic Principles of Plant Quarantine

The basic principles of plant quarantine is to check the entry and spread of potentially dangerous plant pathogens and insects imported along with the germplasm.

4. Goal of Plant Quarantine

The goal of plant quarantine are

- To delay or prevent entry of the pathogen along manmade pathways.

- If entry succeeds, to prevent infection.
- If infection succeeds, to prevent establishment of pathogens entering on manmade and natural pathways.
- If established, to minimize or retard spread.

Pests Introduced in India

Pests	Scientific Name	Year of Intorudction
Lantana weed	*Lantana camara*	1809
Coffee rust	*Hemilia vastatrix*	1879
Late blight of potato	*Phytophthora infestans*	1883
Wooly aphid of apple	*Eriosoma lanigerum*	1889
Downy mildew of Grape	*Plasmopara viticola*	1910
San Jose Scale	*Quadrapidiotus perniciosus*	1911
Water hyacinth	*Eichhornia crassies*	1914
Potato tuber moth	*Phthorimaea operculella*	1937
Banana bunchy top virus		1940
Wart of potato	*Synchytrium endobioticum*	1953
Congress grass	*Parthenium hysterophorus*	1956
Golden nematodes	*Globodera rostochiensis & G pallida*	1961
Apple scab	*Venturia inaequalis*	1978
Downy mildew of sunflower	*Plasmopara halstedii*	1984
Peanut stripe virus		1987
Sunflower necrosis	*Sunflower necrosis virus*	1997
Cotton mealy bug	*Phenococcus solenopsis*	2006
Papaya mealy bug	*Paracoccus marginatus*	2009
Tomato leaf minor	*Tuta abosoluta meyrick*	2014
Root knot nematode	*Meloidogyne enterolobii*	2016
White rust of chrysanthemum	*Puccinia horiana*	2016
Fall armyworm in maize	*Spodoptera frugiperda*	2018

5. Prerequisites of Plant Quarantine Regulations

Plant quarantine regulations have certain prerequisites like

- Only pathogens and pest that pose a threat to major crops or forest should be taken into considerations.
- It should be formulated to control the pests and not to hinder trade or accomplishment of other objectives.
- It should be operated solely under the law.

- Persons involved in quarantine measures should be properly trained and experienced.
- Professional workers and the public must co-operate on an international scale for effective operation of quarantine regulations.

6. Component of Plant Quarantine

Plant protection and quarantine (PPQ) programs usually have three components

- Exclusion of pathogens and pests of quarantine and economic significance.
- Suppression and eradication of exotic pathogens and pests recently introduced along natural and manmade pathways.
- Assistance to exporters of plant products in meeting the quarantine requirements of importing countries.

7. Pest Risk Analysis

Pest risk defines the chances that a pathogen or pest of quarantine significance will enter along a manmade pathway. Risk often is expressed as low, medium or intermediate, and high.

- **Low risk:** Means that there is little chance that the pathogen or pest will enter.
- **High risk:** Means that chances are high that the pathogen or pest could enter.
- **Acceptable risk:** An acceptable risk level means that the benefits derived after taking risk are high enough to justify taking the risk, with safe guards , in the first place.

8. Plant quarantine methods

The plant quarantine activities are:

i) Complete embargoes

- To exclude specified plants and plant products completely from a country infected or infested with highly destructive diseases or pests.
- However, in practice it is difficult to achieve because of more and more exchange of diverse genetic material among countries.

ii) Partial embargoes

- Apply when a pest or disease of quarantine importance to an importing country is known to occur only in well defined area of the exporting country

- An effectively operating internal plant quarantine service exists that is able to contain the pest or disease within this area.

iii) Inspection, treatment and certification at point of origin

- It involves the inspection and treatment of a given commodity when it originates from a country.
- And certification in accordance with quarantine requirements of importing country.

iv) Inspection at the point of entry

- It involves inspection of plant material immediately upon arrival at the prescribed port of entry.
- And if necessary subject to treatment before the same related.

v) Utilization of post entry plant quarantine facilities

- It involves growing of introduced plant propagating material under isolated or confined conditions.

9. Plant quarantine in India

S.No.	National legislations
1	Destructive Insects and Pests (DIP) Act of 1914
2	Plant Quarantine (Regulations of Import into India) Order, 2003
3	Livestock Importation Act, 1898
4	The Environmental (Protection) Act, 1986
5	Biological diversity Act, 2002

- **Destructive Insects and Pests (DIP) Act of 1914** of Central Government, which regulates the introduction of exotic diseases and pests into the country or their spread from one state or union territory to another. The act also empowers the state governments to frame suitable rules and issue notifications for inter-state movement of plant and plant material. Those rules are known as plant quarantine rules.
- **The Agricultural Pests and Diseases Acts** of various states, which suppress or prevent the spread of diseases and pests in areas within a State or union territory.

10. Pesticide Management Bill, 2020

The Pesticides Management Bill, 2020 was approved by the Union Cabinet on 12 February 2020. It replaces the existing Insecticide Act, 1968, which Prakash

Javadekar, the Minister for Information and Broadcasting, mentioned as "age-old and needs immediate rewriting". The aim of the Pesticides Management Bill, 2017 is to mainly

- Regulate the import, manufacture, export, storage, sale, transport, distribution, quality and use of pesticides.
- Control pests.
- Ensure availability of quality pesticides.
- Allow its use only after assessing its efficacy and safety.
- Minimize the contamination of agricultural commodities by pesticide residues.
- Create awareness among users regarding safe and judicious use of pesticides.
- To take necessary measures to continue, restrict or prohibit the use of pesticides with a view to prevent its risk on human beings, animals or environment.

As per the bill, any person, institution or agency shall register any pesticide at the Plant Quarantine and Pesticides Management Centre prior to production, synthesis, import, export, commercial use, storage, sale or distribution, transportation, packing only after the prescribed requirements are met, the centre shall register such pesticides and provide the certificate. If the person, institution or agency does not abide by the terms and conditions referred to in the certificate or if any registered pesticide appears to cause adverse effects on human and animal health, and the environment, the certificate may be suspended or cancelled at any time without assigning reasons whatsoever.

Highlights of the Bill

- The draft bill proposes an increase in penalties on sale of spurious, substandard and misbranded pesticides and gives state governments more power to deal with the issue and take action against them.
- The draft contains clauses relating to allowance of provisional registration of new pesticides in India in case of "national exigency" for a period of 2 years.
- The maximum punishment for violation (such as sale of prohibited or spurious pesticides) is a penalty of up to INR 50 lakh and up to 5 years imprisonment. Earlier, the punishment was limited to INR 2,000 and up to 3 years imprisonment.
- The draft provides for punishment to anyone who 'uses' a pesticide in contravention to the provisions of this Act.

- The proposed bill provides for paying of compensation to the affected farmers or users under the provisions of Consumer Protection Act, 1986.
- The bill contains detailed clauses for registration of new molecules.
- It has also tightened the guidelines for registration and licensing of new molecules.
- It includes a broader category of offences and provisions for paying compensation to the farmers.
- The state government has to report all cases of poisoning to the centre on a quarterly basis according to the draft.
- States can ban chemical pesticides for up to 6 months, unlike the current provision which allows ban for up to 2 month packing and prohibits the same of pesticides which are not registered with the centre.

11. Agencies involved in plant quarantine

Plant Protectionn Network of Directorate of Plant Protection, Quarantine & Storage (DPPQ & S) of India implement the quarantine rules and regulations framed under DIP Act. This organization handles bulk import and export of seed and planting material for commerical purpose.

a) National Plant Protection Organization (NPPO) of India-Ministry of Agriculture & Farmers Welfare, Government of India, New Delhi.

b) Directorate of Plant Protection, Quarantine & Storage (DPPQ & S), Head Quarter-Faridabad

c) Regional Plant Quarantine Station-07

d) Plant Quarantine Stations-59

e) Central IPM Centers-12

f) Central Insecticide lab-01

g) Regional Pesticide Testing Laboratories-02

h) Seaports-34

i) Air ports-12

j) Land custom stations-14

k) Foreign post office-11

12. Domestic Quarantine

The solitary object of enforcing domestic quarantine is to prevent the spread of these diseases from infected to non-infected areas. Presently, domestic plant quarantine exists in following diseases and pests

Pest/Diseases	Host plant material	Restricted States From	To
Wart (*Synchytrium endobioticum*)	Potato	Darjelling West Bengal	Any other state or place in India
Bunchy top (virus) of banana	Banana planting material	Assam, Kerala, Orissa, Tamil Nadu &West Bengal	Any other state or place in India &Uninon Territory
Mosaic (virus) of banana	Banana plants & planting material	Maharashtra & Gujarat	Any other state or place in India &Uninon Territory
Apple scab (*Venturia inaequalis*)	Apple planting materials	J&K and Himachal Pradesh	Any other states of India
Potato Cyst Nematodes (*Globodera rostochiensis* & G pallida	Potato	Tamil Nadu	Any other state or place in India &Uninon Territory
Codling moth (*Carpocapsa pomenella*)	Apple and Walnut plants including fruits	Ladakh districts	Any other area of Jammu & Kashmir
Fluted Scale (*Icerya purchase*)	Many host plant	Mysore (Karnataka), Madras (Tamil Nadu) & Kerala	Any other state or place in India
Coffee berry borer (*Hypothenamus hampei*)	Coffee seeds/plants/powder	Nilgiri (TN), Kodagu (Karnataka) & Wynad (Kerala)	Any other states of India
San Jose Scale (*Aspidiotus perniciosus*)	Many host plant species	Punjab, UP, Madras (Tamil Nadu) & Kerala,WB,Assa m, Orissa, HP, J&K	All other part of India

Questions

1. What is plant quarantine? Describe domestic quarantine measures in India and how they are helpful in crop disease management.
2. Classify the plant quarantine. State the causes of domestic plant quarantine.
3. In which way are plant quarantine measures helpful in the management of plant diseases? Do you justify quarantine measures?
4. Write down the principles and goals of plant quarantine.Give the factors affecting the efficacy of plant quarantines.
5. What are the two different categories of plant quarantine? Give some examples of the plants and plant materials prohibited from entry into India.
6. Why are quarantine measures called legislative measures?
7. Write down the methods of plant quarantine.

8. Write short notes on

i) Plant biosecurity

ii) Domestic quarantine

iii) Agency involved in plant quarantine

iv) DIP act.

28

Biological Management

Definition

Garret (1970) defined *biological control as any condition under which or practice whereby survival or activity of a pathogen is reduced through the agency of any other living organism (except man) with the result that there is a reduction in the incidence of the disease caused by that pathogen.*

Baker and Cook (1974) stated that the *biological control is the reduction of inoculum density or disease producing activities of a pathogen or parasite in its active or dormant stage, by one or more organisms, accomplished naturally or through manipulation of the environment, host or antagonists, or by mass introduction of one or more antagonists.*

Biocontrol can be simply defined as the application of living organism to control diseases.

Types of biological control

1. **Partial Biological Control:** If control is achieved at somewhat higher average population densities than it is economically satisfactory, this is sometimes alluded to as partial biological control.
2. **Substantial Biological Control:** If biological control can be substantially successful economically for a period of time, the only partially so far a subsequent period. Here a degree of biological control is evaluated on economic yard stick.
3. **Complete Biological Control:** If biological agent is successful economically for a very long period of time. So far as such success has not yet been achieved.

Scope of Biological Control

1. Promoting basic research
2. To exploit the bioagent on crop pest
3. Help to reduce pollution hazards

4. Necessity to intensify new horizons of biological control
5. Adoption of biological control methods in agro industries

Mechanism of bio control

Major mechanism of biocontrol are

1. Competition
2. Hyperparasitism
3. Antibiosis
4. Antagonism
5. Exploitation
6. Induction of host resistance
7. Hypovirulence

1. **Competition:** Competition could involve all kinds of *interplay between organisms in which one is favored at the expense of the other.* But in strict sense, if we keep antibiosis or even exploitation restricted to their specific mode of action and results, competition has been defined as a more or less active demand in excess of the immediate supply of material or condition on the part of two or more organism.
2. **Antibiosis:** Antibiosis is defined as the condition in which *one or more metabolites excreted by an organism have harmful effects on one or more other organism*. In such antagonistic relationship species A produces a chemical substance that is harm or inimical to species B without species A deriving any direct benefit. However, the species A may have an indirect benefit in having a better competitive ability, thereby getting an advantage over species B for substrate colonization.
3. **Mycoparasitism:** Mycoparasitism is an act where *one fungus parasitizes the other one*. The mycoparasitism is of common occurrence and examples can be found among all groups of fungi from chytrids to the higher basidiomycetes. It includes different kinds of interaction, viz. coiling of hyphae, penetration, production of haustoria and lysis of hyphae.
4. **Antagonism:** Antagonism implies that in any association of two or more species, at *least one of the interacting species is harmed due to the activities of one or more of the rest.*
5. **Exploitation:** When *species A inflicts harm by the direct use of species B for its own benefit*, it is exploitation (parasitism and predation). The two terms *parasitism* and *predation* have basically same effects, i.e., destruction of the host or prey. In parasitism, some sort of etiological relationship between the parasite and the host is established and the host is not rapidly eliminated. A predator physically eliminates its prey by direct feeding on it without establishing any etiological relationship.

6. **Hypovirulence**: Phenomenon of reduced virulence of a pathogen strain than normal one developed as a result of its infection by ds RNA.
7. **Induction of host resistance**: Plants respond to a variety of chemical stimuli produced by soil-and plant-associated microbes. Such stimuli can either induce or condition plant host defenses through biochemical changes that enhance resistance against subsequent infection by a variety of pathogens

Some examples of mechanism of biocontrol

S.No.	Mechanism	Examples
1	Hyperparasitism/Mycoparasitism	• *Trichoderma harzianum* parasitize mycelia of *Rhizoctonia and Sclerotium*
2	Antibiosis	• HCN produced by *Pseudomonas fluorescens* • Bacilysin produced by *Bacillus subtilis* • Trichodermin produced by *Trichoderma*
3	Competition	• Exudates/leachates consumption· Siderophore scavenging • Physical niche occupation
4	Induction of host resistance	• ISR mediated resistance in tobacco, aridiopsis • Rhizobacteria applied to seeds or roots induce systemic resistance response expressed against pathogens infecting aerial tissues
5	Hypovirulence	• Hypovirulence of fungal pathogens by mycovirus

Role of Biocontrol

1. Disease control
2. Plant growth promoter
3. Stimulation of plant resistance and plant defense mechanism
4. Transgenic plants
5. Bioremediation

Advantages of biocontrol

- Perform preventive control
- Low persistence and residual effect
- Mostly biodegradable and self perpetuating
- Delayed knock down effect
- Less resistance prone

- Less pest resurgence
- Less harmful on beneficial microorganisms
- Mostly target specific
- Waiting period almost nil
- Costlier but reduced number of applications required

Limitations

- Lack of faith and inconsistent field performance
- Poor quality and shelf life
- Competition with chemical pesticides
- High budget of production and lesser agribusiness
- Registration process is expensive and also time-consuming
- Lack of knowledge, confidence and awareness in end user
- Inclination towards use of chemical pesticides
- Non availability of popular literature
- Limited techniques for indigenous and exotic mass production of biotic agents
- Laboratory evaluation of biotic agents and their field efficiency
- Transfer of technology requires more attention

Difference between biopesticide and chemical pesticide

Biopesticide	Pesticides
These do not harm non target species They do not pollute the environment	Nontarget species are also harmed Cause pollution; some times serious
No harmful residues remain in food, fodder and fibers	Harmful residues may often remain in food, fodder and fibers
Relatively cheaper	Relatively costlier
Pests are expected not to develop resistance to biopesticides	Pests may become resistant, e.g., Heliothis has become resistant to most insecticides
Since they are highly specific, correct identi fication of the pest is essential	It is often not critical
High specificity may often make the use of two or more biopesticides necessary	Often not required
Performance may be variable due to the influence of biotic and abiotic factors of the environment	This is not often the case

Most commonly used biocontrol agents (BCAs)

The most commonly used biocontrol agents for control of plant pathogens are:

1. Trichoderma
2. Plant Growth Promoting Bacteria (PGPR)

What is *Trichoderma*?

Trichoderma is the most common BCAs used in plant disease management. The most widely used Trichoderma genus are strains of *T. viride*, *T. virens* and *T. harzianum*. Several plant diseases collar rot, damping off, wilt, rhizome rot, sheath blight caused by fungi can be potentially controlled by *Trichoderma* species.

Secret of success of *Trichoderma* as BCAs

The success of *Trichoderma* strains as BCAs is due to their

- High reproductive capacity.
- Ability to survive under very unfavorable conditions.
- Efficiency in the utilization of nutrient.
- Capacity to modify the rhizosphere
- Strong aggressiveness against phyto pathogenic fungi
- And efficiency in promoting plant growth and defense mechanisms.

Plant Growth Promoting Rhizobacteria (PGPR)

The bacteria which promote the plant growth are called PGPR. PGPR, fall under genera *Pseudomonas, Bacillus, Arthrobactor, Achromobacter, Citrobactor, Enterobacter* and *Flavobacterium*. The PGPR besides enhancing growth and yields, are also potential biocontrol agents.

Ways that PGPR promote plant growth

- Increasing nitrogen fixation in legumes.
- Increasing supply of other nutrients, such as phosphorus, sulphur, iron and copper.
- Producing plant hormones.
- Enhancing other beneficial bacteria or fungi.
- Controlling fungal and bacterial diseases.

Method of Application of *Trichodema*

Biocontrol agents are applied through

- Seed treatment
- Soil application

Seed treatment

- Treat seed with formulation of Trichoderma or Pseudomonas @ 5-10 gm/kg of seed depending upon the quality of product.

Soil application

- Formulations of *Trichoderma* or Pseudomonas are used as soil application by mixing in well decomposed compost before application into soil.
- The quantity required varies from 125-250 kg/ha. It is mixed with, FYM and broadcast on the soil and then incorporated into soil through harrowing.
- It is used to control *Fusarium* wilt, *Sclerotium* foot rots and *Macrophomina* root rots.
- Applied at least 2 weeks before sowing of crops.

Commercially available BCAs

Table- Commercially available BCAs, their trade name (s), and the target

Product/Trade name	Species/strain of *Trichoderma*	Agency/Company
Ecofit, Trichogourd, Defense SF, Tricho-X, Biogourd	*Trichoderma viride*	*Fusarium, Rhizoctonia, Pythium, Phytophthora,Nectria*
Top shield ,Root shield, F-Stop, Trichodex	*Trichoderma harzianum*	*Fusarium, Rhizoctonia, Pythium, Pythium Colletotrichum, Monilinia, Plamopara , Rhizopus, Sclerotinia*
Bioderma	*Trichoderma viride+ Trichoderma harzianum*	*Fusarium, Botryosphaeria, Fusarium*
Ecoderma	*Trichoderma viride+ Trichoderma harzianum*	*Rhizoctonia, Pythium, Phytophthora*
Binap- T&W	*Trichoderma harzianum+ T. polysporum*	Wood decay fungi
Gilogard and Soil guard	*Trichoderma virens*	*Rhizoctonia, Pythium*
AQ 10 biofungicide	*Ampelomyces quisqualis*	Powdery mildew
Biotrox C	*Fusarium oxysporum* (non pathogenic)	*Fusarium oxysporum*
Galltrol	*Agrobacterium radiobactor* strain 84	*A.tumefaciens*
Companion	*Bacillus subtilis strain GB03*	*Pythium, Phytophthora, Fusarium, Rhizoctonia*
Biosave 10LP, 110	*Pseudomonas syringae*	*Botrytis, Mucor, Penicillium*
Dagger G	*Pseudomonas fluorescens*	*Rhizoctonia , Pythium*

Questions

1. Define biocontrol. Discuss its scope and limitations to control plant pathogens.
2. Write an essay on biological control of plant diseases.
3. What is biological control of plant pathogens? Give some suitable examples where it has been successfully used to control plant diseases.
4. What is a suppressive soil?

5. What are siderophores? How siderophores production by biological control agnets affects plant pathogen.
6. What do you understand by mycoparasitism? Give some suitable examples of mycoparasitism.
7. Differentiate between chemical and biological control.
8. Mantion the commercially available fungal biocntorl agents at present along with their trade name.
9. Write down advantage and diasdvantage of biological control.
10. What is biological control?. Discuss various mode of action of biological control agents.
11. Give the characteristics of an effective biological control agent? How BCAs are screened and developed?
12. Attempts the following :

a) Suppressive soils.

b) Siderophores

c) Antibiosis

d) Hypovirulence

e) Future of biological control

13. How do you understand by biopesticides? Give name, dose and disease managmed by three most commonly used biopesticides.'

29

Host Plant Resistance

Definition

Those characters that enable a plant to avoid tolerate or recover from attacks of pathogens under conditions that would cause greater injury to other plants of the same species.

OR

The *inherent ability* of an *organism* (i.e., *the crop* plant) to *resist* or *withstand the pathogen* is called resistance

Classification of Resistance

Each plant species is affected by hundreds of kinds of pathogen. Yet, those plants survive which are resistant, adapting a number of mechanisms classified variously.

1. Based on existence

a) Preformed

When the resistance is already present in the plant even the absence of the pathogen. it is known as preformed, axenic or passive resistance. Example, mildew resistance in barley variety Nigrete.

b) Induced

When the resistance is not present in the plant in the absence of the pathogen , but with its contact the resistance is induced, it is known as active, apergic or induced resistance.

Example, seedlings of cotton are susceptible to many pests due to absence of gossypol but once infected by *Verticillium, gossypol* is synthesized by induction and seedlings become resistant.

2. Based on type of host response

a) Immune:*Immunity is exempt from infection or hundred per cent freedom from disease*. No symptom develop. For instance, potato is resistance to karnal bunt.

b) Resistance: *Resistance characterizes those situations in which some degree of host pathogen interaction is evoked.* Therefore, resistance is partial, that is ,always some symptoms appear contrary to immunity, where no symptoms appear.

c) Tolerance: *Tolerance is defined as the inherent or acquired capacity of the host to endure disease.* Tolerance means disease tolerance where yield losses are minimum even in the presence of disease.

2. Based on epidemiological terms

a) Vertical resistance

When a plant variety is more resistant to a few races of pathogen than to others, the resistance is called 'vertical' or 'perpendicular'.

b) Horizontal resistance

When host resistance is equally effective against all races of a pathogen-it is termed horizontal or 'lateral'.

Table 25: Major difference between horizontal & vertical resistance

Horizontal	Vertical
Minor	Major
Polygenic	Monogenic
Race nonspecific	Race specific
Quantitative	Qualitative
Multiple gene	Multiple allele
Durable	Nondurable
Non hypersensitive	Hypersensitive
Low to moderate	High
General resistance	Specific resistance

Disease Escape or Apparent resistance

It refers *to the freedom of susceptible host plant varieties from a disease purely due to the environmental factors.* It occurs primarily due to avoiding contact and also by preventing infection.

It may be achieved by

- Early varieties
- Change date of planting
- Change in the site of planting
- Balanced application of NPK

- Control of disease carrier
- Example: Early maturing or short duration variety may escape serious disease attack (wheat rust and potato late blight)

Gene for Gene hypothesis

According to Gene for Gene hypothesis *for every resistance gene present in the host, the pathogen has a gene for virulence.* Susceptible reaction would result only when the pathogen is able to match all the resistance genes present in the host with appropriate virulence gene. In most of the pathogens, virulence is recessive to avirulence. The hypothesis is proposed by *HH Flor* in linseed rust (*Melampsora lini*)

Mechanism

1. Compatible reaction
2. Incompatible reaction

1. Compatible reaction

- The alleles for susceptibility (r) in the host and those for virulence (a) in the pathogen produce specific compounds, which interact with each other to produce the susceptible response.
- A lack of interaction between the products of r and a genes specifiy resistant response of the host.
- It is found in heterotrophic pathogens.

2. Incompatible reaction

- This type of gene for gene relationship is found in biotrophic pathogens.
- The alleles for resistance in the host (R) and those of avirulence in the pathogen (A) produce specific compounds which recognize each other , when these specific compounds interact , they produce resistance response in the host
- The lack of interaction between the products of the genes for resistance and avirulence produces susceptible reaction in the host.

	Virulence gene	
Resistance Gene	A_1	a_1
R_1	R	S
r_1	S	S

- Incompatible reaction is associated usually with hypersensitive responses of the host.

Mechanism of Disease Resistance

1. Mechanical
2. Hypersensitivity
3 Nutritional

1. **Mechanical**: Mechanical or anatomical features of the host may prevent infection e.g. Closed flowering habit of wheat and barley prevents infection from the spores of ovary infecting fungi.
2. **Hypersensitivity**: Immune reaction like localized cell death found in case of biotrophic organisms. e.g. yellow rust.
3. **Nutritional**: The reduction in growth and spore production is generally suppressed due to unfavourable physiological conditions within the host.

Source of disease resistance

- A known variety
- Germplasm collection
- Related species
- Mutations
- Somaclonal variation
- Unrelated organisms

Method of breeding for development of resistance

The methods of breeding varieties resistant to diseases do not differ greatly from those adopted for other characters. The following methods are used:

1. Introduction
2. Selection
3. Hybridization followed by selection
4. Mutation
5. Somaclonal variation
6. Genetic engineering

Strategies against viral pathogen

- Coat protein virus-most successful
- Sat-RNA
- Defective viral gneome
- Antisense construct of critical viral genes
- Ribozymes

Strategies against fungal and bacterial pathogen

By expression of the following transgens:

- Genes encoding insensitive target enzymes
- Genes specifying toxin inactivation
- Expression of antibacterial peptides
- Expression of bacterial lysozymes
- Gene specifying artificial programmed cell death
- Expression of phytoalexins
- Gnens encodying ribozyme inactivation proteins
- Expression of heterologous thionins

Resistance Breeding

In resistance breeding, one will have to consider the biology of the of pathogen & pest over and above the breeding system of crop plants and the types and genetics of resistance. The use of specific resistance demands the identification and use of strong genes against which the stabilizing selection operates.

There are various ways in which strong gene can be used.

- Singly
- Deployment
- In combination
- Multiline

a) Use of single resistance gene

The single gene is used one at a time. This strategy results in homogeneity of the crop, which imposes strong selection forces on pathogen population. The use of single R-gene is thus risky and its use has been questioned on the ground of its failure to provide permanent resistance.

b) Gene Deployment

It refers to *strategic use of major gene, by deploying these over space and time to avoid phenomena of boom and bust.*

Advantages

- It helps in preventing disease
- Maintain diversity of major genes in different geographical region.

Drawback

- Difficult to find out different genes with similar effectiveness.

The gene deployment strategy has been used for controlling many plant diseases such as bacterial blight of rice, rusts of wheat, crown rust of oat, leaf rust of barley, rust of bean etc.

c) Pyramiding of genes

When two or more known resistance genes are introduced into a single variety it is called pyramiding or stacking of gene.

Advantages

- Provide resistance to a common spectrum of races.
- Provides supplementary resistance by increasing breadth, not the depth of resistance.
- Provides durable resistance used in oat against crown rust (Cavada).

Disadvantages

- Chances of development of super race
- Tedious
- Agronomic performance of cultivar restircted

d) Multilines

Multiline varieties are mixtures of several purelines of similar height, flowering and maturity dates, seed color and agronomic characteristic, each of which has a different gene for resistance to the given disease. Multiline variety appears to be useful approach to control diseases like rusts where new races are continuously produced. In India, three multiline varieties viz., KSML3, MLKS 11 and KML 7406 have been released in wheat

Boom and Burst Cycle

When as new variety carrying vertical resistance is developed it is resistant to the prevalent pathotype of the pathogen in question, therefore it becomes popular rapidly and occupy a large area under the concerned crop this period of boom continues till the pathotype specific to the VR gene employed in the variety evolves and or increase in frequency to cause an epidemic in the fields occupied by the variety. Now susceptible variety loss popularity and area under it declines sharply thus the variety is effectively busted by the virulent pathotype. This has been called as *boom and bust cycle* by Priestley. The boom burst cycle can be broken by

- Localized usage of R-gene
- Cycling of R-gene
- Use of polygenic resistance
- Combination of above
- Multigenic resistance
- Multiline variety

Vertifolia effect

The process of neglecting or losing horizontal resistance in the course of breeding for vertical resistance has been termed as 'vertifolia effect'. This failure occurs because of two reasons:

- The level of horizontal resistance in the varieties carrying oligogenes for resistance is usually low.
- The pathogen is able to evolve the virulent pathotype.

Pre-emptive breeding

Pre-emptive or anticipatory breeding for resistance is breeding for resistance to *future patho types.* Its success depends on the ability of the breeder to predict the likely pathogen phenotypes (pathopypes) that will be important at some future time.

Questions

Short Answer

1. What is durable resistance? Give five examples of durable resistance.
2. Name two diseases in which the soil moisture determines disease escape.
3. Discuss about

i) Gene pyramiding

ii) Gene deployment

iii) Apparent resistance

4. Differetiate between horizontal and vertical resistance.
5. Write down the mechanism of disease resistance.
6. What are the disadvantages of growing genetically uniform crops over large areas.
7. Differentiate between disease escape and tolerance.
8. Name four diseases (with their pathogens) that exemplify vertical resistance.
9. Define pathotype,pathodeme,Esodemic and Exodemic.
10. What are Hrp proteins?
11. Write in brief about cross protection

Long Answer

1. What is host plant resistance? Classify resistance on the basis of existence, host response and epidemiological competence.
2. Write short notes on
   a) Boom and Bust cycle
   b) Vertifolia effect
   c) Anticipatory breeding
   d) Multiline
3. Who first proposed 'gene-for-gene concept'? Name the disease involved with its pathogens.
4. What are resistance genes? Classify them on the basis of their structure and function. Discuss their mechanism of action and evoloution.
5. Give a concise account of genetic resistance in plants.
6. Attempt the following:
   a) Boom and bust cycle
   b) Genetic recombination in viruses
   c) Cytoplasmic resistance
   d) Differences between VR and HR
   e) Physiological races
   f) Avirulence genes

30

Chemical Management

Introduction and Concept

The word fungicide has originated from two Latin words: viz. *fungus and caedo (to kill).* So literally, a fungicide would be any physical or chemical agency which kills a fungus. However, the word is restricted to chemicals. Hence, the word *fungicide* should mean a *chemical capable of killing fungi.* However, there are a number of compounds which do not kill the fungus. They simply inhibit growth or spore germination temporarily. If the fungus is freed from such substances, it would revive. Such a chemical is called a *fungistat.* and the phenomenon as *fungistasis*. Some other chemicals, like certain phenantherene derivatives and bordeaux mixture may inhibit spore production without affecting vegetative growth. These are called *antisporulants*. There are other groups of chemicals which exhibit very poor or no antifungal activity *in vitro* condition but provide protection to the plants either by inhibiting the penetration of host surface by the fungi or by inducing the host defense system. The former type of chemicals are termed as *antipenetrants* and later as *antipathic agents*. Even though fungistats, antisporulants, antipenentrants, and antipathic agents do not kill fungi, they are included under the broad term fungicide because by common usage, the word *fungicide has been defined as chemical substance which has ability to prevent damage caused by fungi to plants and their products*. Fungicide which is effective only if applied prior to fungal infection is called *protectants*. On the other hand, fungicide which is capable of eradicating a fungus after it has caused infection, and thereby curing the plant, is called *therapeutant*.

Characters of chemicals or fungicides

In general, the anti-infection chemicals or pesticides having following characters are supposed to be ideal.

- High field performance.
- Easy availability of the active constituents.
- Absence of toxicity for the host, man and animals.

- High toxicity for the pathogen at low concentration.
- Retention of toxicity on dilution.
- Stability in storage.
- Slow or no loss of toxicity in storage.
- Good spreading quality on host surface.
- High tenacity on the host surface.

Advantages and disadvantages of chemicals or fungicides

S.No.	Advantages	Disadvantages
1	Quick-acting	Potentially toxic to humans
2	Easy to use	Impact on non-targets
3	Cheap (relatively)	Potential negative environmental effects
4	Readily available	Unknown cumulative effects
5	Can save lives (vector-borne diseases)	Pest resistance

Classification of Chemicals

The chemicals can be broadly classified on the basis of their

- Mode of action against pathogen
- Type of pathogen

A. According to mode of action

Depending upon their mode of action the chemicals can be grouped as protectants, eradicants and therapeutants.

1. Protectants

- Protectants are prophylactic in their behaviour.
- They act outside the plant parts as a cover to check the invasion by the pathogen.
- They include thiram, captan, agallol, zineb, sulphur, ceresan and streptocyclin.

2. Eradicants

- Eradicants help to eradicate the dormant or active pathogen from the host completely.
- These chemicals can be used as protectant as well as eradicant.
- Examples are lime, sulphur, organo mercurials.

3. Therapeutants

- Therapeutants is an agent that inhibits the growth and development of a disease already entered in a plant, when applied appropriately.
- Usually the chemo-therapeutants are systemic in their action, i.e. they enter the plants and affect deep-seated infection.
- E.g. plantvax, vitavax, dithane M-45, streptocyclin and agromycin.

B. According to type of pathogen

Based on the type of pathogens, the chemicals can be grouped as

1. Fungicides
2. Bactericides
3. Nematicides

1. Fungicides

- Chemicals which kill fungus are called fungicides.
- These fungicides can again be classified as *contact* and *systemic* based upon their action in plant.

Properties of contact and systemic fungicides

S.No.	Contact Fungicides	Systemic Fungicides
1	Multisite inhibitors	Single site inhibitors
2	Protective in function	Curative in function
3	Many metabolic systems of fungal pathogens affected	Few metabolic systems of fungal pathogens affected
4	Phytotoxicity is common	Phytotoxicity is rare
5	Target non specific	Target specific
6	Pathogen resistance development rarely takes place	Pathogen resistance development commonly takes place
7	They are localized in action	They are systemic in action
8	They are confined to redistribution on plant surfaces	They are translocated inside the plant

I. Contact fungicides

- Are the chemicals which do not enter inside the plant tissue, but kill the pathogen by surface contact.
- Some of the important categories of chemicals used for disease control are:

A. Sulphur fungicides
B. Copper fungicides
C. Mercury fungicides
D. Quinone fungicides
E. Benzene fungicides
F. Hetero-cyclic compounds
G. Organo phosphate fungicides
H. Soil fumigants

A. Sulphur fungicides

Sulphur fungicides are the *oldest method* of disease control.Several inorganic and organic sulfur fungicides are in use for control of plant diseases. All sulfur fungicides are nonsystemic, and except lime sulphur, are phytotoxic (excluding sulfur shy varieties) and compatible with most of the pesticides. However, the most popular fungicides in sulphur groups are the organic compounds known as *dithio-carbamates*. As far as their mode of action is concerned, elemental sulphur interferes with energy production by intercepting electron on the substrate side of cytochrome C in the mitochondrial electron transport system. The dialkydithiocarbamates are known to inhibit a multitude of enzymes; therefore, fungitoxicity probably involves concurrent inhibition of enzymes at several sites. The pyruvate dehydrogenase reaction is particularly highly sensitive to dialkydithiocarbamates.Some of the important chemicals under dithiocarbamates are thiram, ferbam, ziram, nabam, zineb and maneb.

B. Copper fungicides

Ever since the discovery of Bordeaux mixture in 1885, copper fungicides predominated the field of fungicidal plant disease control for more than 50 years until synthetic organic fungicides invaded the market. Even today some of the copper compounds are used widely in many countries. Copper is a *multisite biochemical inhibitor* (probably interact with-SH groups of enzymes) with little biological specificity.Bordeaux and Burgundy mixtures are the dispersible forms of cuprous oxide and basic carbonates are among the important copper fungicides.

C. Mereury fungicides

Mercury is a general biocide. Several organomercurial compounds were introduced. Most of them are now withdrawn due to high mammalian toxicity. Only two organic and three inorganic mercury compounds are in use and that too in very limited cases. Mercury also exhibits *multisital action* due to its interaction with the–SH group of the suscetible enzymes. Many mercury compounds are highly effective as fungicides and bactericides. Example: Agallol, Aeretan

D. Quinone fungicides

One of the quinone derivatives, chloranil was introduced as a fungicide in 1940. Now it has been replaced by dichlone, another quinone derivative introduced in 1943. Quinone fungicides are *multisital* in their mode of action. The two mechanisms which are thought to be most likely for dichlone and chloranil are 1. Binding of the quinone nucleus to –SH and –NH2 groups in fungal cell, and 2. Interference with electron transport system (Nene and Thapliyal, 1996). Among the Phthalamide compounds, captan was introduced first and is still being used widely. Basically these fungicides are *protective* in nature but limited systemicity has been reported for captan and captafol. Phthalamides are highly reactive against thiol (–SH) groups of proteins (enzymes) and low molecular weight metabolites (cystein, glutathione, etc).

E. Benzene fungicides

Benzene is used as a dormant spray for the control of many diseases of fruit trees and ornamental plants and for the treatment of wounds in trees. Karathane is an excellent substitute for sulphur for the control of powdery mildew. It is also effective against mites. Chloronil is active against *Rhizoctonia solani, Sclerotium rolfsii* and *Phytophthora cinnamoni*. This fungicide has been used for the most part as a seed or in-furrow treatment of cotton.

F. Hetero-cyclic compounds

Hetero-cyclic compound are used as foliage protectants and eradicants of fruits and vegetables. Examples: captan, difoltan and folpet.

G. Organo phosphate fungicides

The organo phosphate include Fosetyl-AI, Kitazin (IBP), edifenphos (Hinosan) and pyrozophos. Fosetyl-AI is very rffective againts floiarm root and stem diseases caused by Phytophthora, Pythium and downy mildews in wide variety of crops. Fosetyl-AI has been reported to stimulate defense reactions and the synthesis of phytoalexins against oomycetes. Kitazin and edifenphos are effective against rice blast and other disease, and pyrozophos, which is effective against powdery mildew, Bipolaris and Drechstera diseases on various crops.

The heterocyclic compounds are include captan, iprodione and vinclozolin. Most of them inhibit production of compound containing -NH, and SH groups. Captan is an excellent fungicide for control of leaf spots, blight and fruit rots. It is also used as a seed protectant and reported to repel "seed pulling birds". Iprodione is a broad spectrum foliage contact fungicide. It inhibits spore germination and mycelial growth but shows mosty preventive and only early curative, activity. It is effective against *Botrytis, Monilinia, Sclerotinia* and *Alternaria.* Vinclozolin is contact protective fungicide effective against *Botrytis, Monilinia, Sclerotinia.*

H. Soil fumigants

The most promising method of controlling nematodes in the field has been through the use of chemicals called nematicides. Example: chloropicrin, methyl bromide, vapam and vorlex.

II. Systemic fungicides

Systemic fungicides are the fungitoxic compounds which when applied on different parts of the plant are absorbed by the plant tissues and then translocated upwards, downwards and both ways and act on the pathogen either directly or through its metabolite products and control plant diseases away from the point of application.

A fungicide when applied on the plant surface either may remain on the surface (non-systemic) or absorbed by the plant (systemic). The latter *may remain there in treated plant parts (locosystemic)*; *may move in the direction of evapotranspiration stream (apoplastic)* or *may move with photosynthates to sink (symplastic)* or *in both directions (ambimobile).* The *symplast* is the living part of the plant which is enclosed by membranes, i.e. protoplasts, and plasmodesmata, including the phloem sieve cells. Long distance transport in phloem is symplastic. The *apoplast* is the nonliving part of the plant, i.e., cell walls and cuticle, including xylem vessels and trachieds. Long distance movement in the transpiration stream is *apoplastic.*

Apoplastic Translocation

Apoplastic translocation, within the plant, is usually directed from roots to transpiring areas, especially leaves. Fungicides are absorbed by the roots, mainly in root hair zone, along with water. Root absorption is usually a passive phenomenon. Radial movement through the cortex zone occurs either symplastically or apoplastically. Symplastically moving substances cross the plasmalemma and are transported via protoplast and plasmodesmata of cortex cells and then through the epidermis cells to the vessels of the xylem.

Symplastic Translocation

Symplastic movement takes place within the living parts of the cell. The movement is in the sieve tubes of the phloem tissues. This is an active movement and requires the expenditure of metabolic energy both for uptake inside the cells as well as for movement in the sieve tubes. Fully expanded photosynthesizing leaves serve as a source and roots, flowers, young growing leaves, and fruits serve as a sink. A fungicide with symplastic movement, after entry into the phloem, follows the same source to sink pathway as followed by the phloem assimilates.

Ideal Systemic Fungicide

An ideal systemic fungicide should have the following characteristics.

1. Systemic fungicides are highly selective i.e. toxic to the pathogen but not to the host.
2. May alter the metabolism of the host so that biochemical or physical resistance to pathogen may be induced or enhanced.
3. Must not adversely affect the host plant to such an extent that the quantity or quality of the crop is reduced.
4. Either be toxic to the pathogen concerned or be converted in the host plant to such a fungitoxicant.
5. Substance must be absorbed sufficiently the translocated from the point of application to the site of the pathogen and should have a considerable degree of stability within the host plant.
6. Most systemic fungicides are translocated in the apoplast, though some appear to move in the symplast also.

On the basis of chemical structure, systemic fungicides can be classified as follows:

1. Oxanthiins
2. Benzimidazoles
3. Sterol Biosynthesis Inhibiting Fungicides
4. Anti-oomycetes fungicides
5. Antipenetrants
6. Host Defense Inducing compounds
7. Miscellaneous systemics
8. Strobilurins fungicides

1. Oxanthiins

Oxanthiins were first systemic fungicides to be discovered and introduced for plant disease control. These fungicides are readily absorbed by the seeds, roots, and leaves and translocated apoplastically. They are very effective mainly against *smuts, bunts* and *rusts* of cereals. Both have low animal toxicity and are quickly degraded. Their mode action is very specific. They interact with succinate ubiquinone reductase complex resulting in *inhibition of oxidation of succinate via electron transport chain*. Examples: *Vita vax* or *carboxin* and *Plant vax* or *oxycarboxin*. Carboxin is used as seed treatment and is effection against damping off and various smut of grain crops. Oxycarboxin is used as seed or boliar treatment and control wide variety of rust disease.

2. Benzimidazoles

Benzimidazoles and thiophanates represent a group of highly effective broad spectrum systemic fungicides. Most *Ascomycetes*, some of the *Basidiomycetes* and *Deuteromycetes*, and none of the *phycomycetes* are sensitive to these fungicides. Their mild cytokinin like effects on some plants tend to retain chlorophyll and in some cases increase yield and delay maturity. Benzimidazoles bind with the Beta- tubulin subunit of microtubules of sensitive fungi and thereby inhibit formation of spindle and subsequently chromosomal separation during nuclear division. Examples: benomyl carbendazim, thiabendazole (TBZ), thiophanate. Carbendazim is a safe broad-spectrum fungicide effective against leaf spots, blotches, blights, scab, seed and soil borne diseases. Thiphanate methyl is a broad spectrum preventive and curative fungicide for use to control powdery and downy mildew, numerous leaf and fruit spots.

3. Sterol Biosynthesis Inhibiting Fungicides

This group includes numerous fungicides from chemically heterogenous classes. However, all these compounds inhibit *ergosterol biosynthesis* in fungi and therefore, are classified as ergosterol biosynthesis inhibitors (EBIs) or sterol biosynthesis inhibitors(SBIs). They are effective against a wide range of fungi except oomycetes which do not need sterol for growth. Several fungicides belonging to morpholines, piperidine, piperazine, pyridine, and azole are known sterol biosynthesis inhibitors. The triazoles includes triadimefon, triadimenol, bitertanol. propiconazole, ectaconazole and difenoconazole. They show long protective and curative activity against a broad spectrum of foliar, root, leaf spot, blight, powdery mildews rusts, smuts etc. They are applied as foliar sprays, seed and soil treatments.

4. Anti-Oomycetes Fungicides

The past decade has witnessed the introduction of five classes of fungicides controlling diseases caused by oomycetes:

1. Carbamates (e.g. prothiocarb, propamocarb)

2. Isoxazoles (e.g. hymexazol)
3. Cyanoacetamide oximes (e.g. cymoxanil)
4. Ethyl phosphonates (e.g. fosetyl-AL)
5. Phenylamides: It cover three groups of compounds: I. acylalanines (e.g. furalaxyl, metalaxyl, benelaxyl), II. Acylamino butyrolactones (e.g. ofurace), and III. Acylamino-oxazo lidinones (e.g. cyprofuram, oxadixyl).

5. Antipenetrants

There are several compounds which show no or poor fungitoxicity under *in-vitro* conditions but prevent infection by interfering with the penetration process.

6. Host Defense Inducing compounds

Various chemicals which have been found to trigger host resistance mechanism and by doing so provide protection against diseases on prior or sometimes simultaneous application with the pathogen. They include both chemical constituents of plants or microbes (biomolecules or metabolites) and synthetic compounds which are foreign to the living system (xenobiotics).

7. Miscellaneous systemics

There are miscellaneous fungicides belonging to diverse chemical groups which show at least some degree of selectivity in their antifungal spectrum. Several excellent systemic fungicides of different chemical composition and affiliation are included in the miscellaneous category. Example- chloroneb, ethozol imazalil, prochloraz, dodine.

8. Strobilurins fungicides

Strobilurins are a group of chemical compounds used extracted from the fungus *Strobilurus tenacellus* and hence the name strobilurin. Also called *QoI inhibitor fungicide.* It control wide range of plant diseases caused by all major groups of fungi, Asco, Basidio, and the Oomycota. Some common Strobilurins are azoxystrobin, kresoxim-methyl, picoxystrobin, fluoxastrobin, oryzastrobin, dimoxystrobin, pyraclostrobin and trifloxystrobin. Strobilurins are miracle fungicides ever developed.

Modes of Action of Fungicides

Application of chemicals to plants in order to prevent or inhibit disease development is a fundamental means of managing diseases caused by fungi. Knowledge of the effectiveness of particular compounds is important for achieving effective disease control. Equally important is an understanding of the underlying physiological mode of action of plant disease management materials.

Fungicides are metabolic inhibitors and their modes of action can be classified into four broad groups.

1. Inhibitors of electron transport chain.
2. Inhibitors of enzymes.
3. Inhibitors of nucleic acid metabolism and protein synthesis.
4. Inhibitors of sterol synthesis.

1. Inhibition of electron transport chain (Respiration in mitochondria)

- **Sulfur:** Disrupts electron transport along the cytochromes
- **Strobilurins:** (azoxystrobin, kresoxim-methyl, pyraclostrobin, trifloxystrobin) Inhibit mitochondrial respiration, blocking the cytochrome bc1 complex.

2. Inhibition of enzymes

- **Copper:** Nonspecific denaturation of proteins and enzymes.
- **Dithiocarbamates:** (Maneb, Zineb, Mancozeb, etc) Inactivate –SH groups in amino acids, proteins and enzymes.
- **Substituted aromatics:** (Chlorothalonil, PCNB) Inactivate amino acids, proteins and enzymes by combining with amino and thiol groups.
- **Organophosphonate (Fosetyl-Al):** Disrupts amino acid metabolism.

3. Inhibition of nucleic acid metabolism and protein synthesis

- **Benzimidazoles:** (Thiophanate-methyl, Carbendazim, Benomyl) Inhibit DNA synthesis (nuclear division).
- **Phenylamides:** (Metalaxyl, Mefenoxam) Inhibits RNA synthesis.
- **Dicarboximides**: (Iprodione, Vinclozolin) Inhibits DNA and RNA synthesis, cell division and cellular metabolism.

4. Inhibition of sterol synthesis

- Imidazoles (imazalil)
- Triazoles (Propiconazole, Myclobutanil, Tebuconazole, Triflumazole)
- Morpholines (Dimethomorph)
- Inhibits sterol production at different site than imidazoles and triazoles. Affects cell wall production.

Table. Mode of action, targetsite and resistance risk attached to different chemical classes of fungicides

Fungicide class	Mode of action	Target site	Resistance risk
Anilino- pyrimidines	Aminoacids and protein synthesis	Methionine biosynthesis	Moderate
2-Aminopyrimidines	Nucleic acid synthesis	Adenosin- deaminase	Moderate
Benzimidazoles	Mitosis and cell divison	B-tubulin assembly	High
Carbamates	Lipids and membrane synthesis, fatty acid	Cell membrane	Moderate
Cinnamic acid derivatives	Glucan and cell wall synthesis	Cell wall synthesis	Moderate
Cyano-acetamide oximes	Nucleic acid/protein synthesis	Undefined	Moderate
Copper compounds, inorganic sulphur, dithiocarbamates, phthalimides, chlorophenols	Multisite disruption of cell function	Reaction with enzymes having –SH amino and hydroxyl groups	Low
Carboxamides	Nucleic acid synthesis	Complex II : succinate dehydrogenase	Moderate
Demethylation inhibitors (DMIs)	Sterol biosynthesis inhibitors	C14- demethylase in sterol biosynthesis	Moderate
Dicarboxamides	Lipid and membrane synthesis	NADPH cytochrome c-reductase in lipid peroxidation	High
Morpholins	Sterol biosynthesis inhibitors	D 8-7 isomerase in sterol biosynthesis	Moderate
N-phenylcarbamates	Mitosis and cell divison	B – tubulin assembly	High
Phenylamides	Nucleic acid synthesis	RNA polymerase	High
Phenylpyrroles	Signal transduction	MAP proetein kinase	Moderate
Strobilurins (Q_oI) fungicides	Respiration	Cytochrome III: cytochrome be I(ubiquinol oxidase)	High

Chemical Combinations

Combination products or Pre – combination mixes are products constituted by mixing two or more compatible fungicide molecules in the right concentration. Proportion of one component is relatively higher than the other. Diseases to be controlled should be susceptible to each fungicide used in mixture. The reasons for combining fungicides are:

- To control different types of disease with a single application.
- To provide superior control than that obtained from one fungicide.
- To increase the effectiveness of one of the chemicals (Synergism).

Table 1: Formulation with combination of fungicides

S.No.	Combination Product	Company
	A. FUNGICIDES	
1.	Azoxystrobin 11.0% + Tebuconazole 18.3% SC	M/s GSP Crop Science Pvt. Ltd.,
2.	Azoxystrobin 7.1 % + Propiconazole 11.9% SE	M/s Adama Makhteshim Ltd..
3.	Boscalid 25.2+ Pyraclostrobin 12.8% WG	M/s BASF India Ltd., Mumbai
4.	Carbendazim 12% + Mancozeb 63% WP	M/s United Phosphorus Ltd., Mumbai
5.	Carbendazim 25% + Mancozeb 50%WS	M/s Indofil Industries Ltd.
6.	Carbendazim 25% + Flusilazole 12.5% SE	M/s Dhanuka Agritech Ltd.
7.	Carboxin 17.5% + Thiram 17.5% FF	M/s Chemchura Chemical Pvt. Ltd.
8.	Carboxin 37.5% + Thiram 37.5% WS	M/s Crompton Specialaties Asia Pacific.
9.	Dimethomorph 12% + Pyraclostrobin 6.7% WG (FI)	M/s. BASF India Pvt. Ltd., Mumbai
10.	Famoxadone 16.6% + Cymoxanil 22.1% SC	M/s EI Dupont India Pvt. Ltd, Gurgaon
11.	Fenamidone 10% + Mancozeb 50% WDG (FI)	M/s Bayer Crop Science Ltd
12.	Fluopyram 17.7 % + Tebuconazole 17.7% SC	M/s Bayer Crop Science Ltd., Mumbai
13.	Hexaconazole 4% + Zineb 68% WP	M/s Indofil Industries Ltd.
14.	Iprodione 25% + Carbendazim 25% WP	M/s Aventis Crop Science Ltd., Mumbai
15.	Kasugamycin 5% + Copper Oxychloride 45% WP	M/s. Dhanuka Agritech Ltd.
	B. INSECTICIDE + FUNGICIDE	
1.	Flubendiamide 3.5% + Hex-aconzole 5% WG	M/s Rallis India Ltd.,

Antibiotics

Defination

"Antibiotic is defined as a chemical substance produced by one microorganism, which in low concentration can inhibit or even kill other microorganism".

Character of good antibiotics

- It should be broad spectrum.
- It should be safe to plants.

- It control plant diseases by acting on the host or on the pathogen.
- It should be active on or inside of the plant.
- It tolerates oxidation, UV irradiation, rainfall and high temperature.
- It should be absorbed and translocated systemically by a plant.
- Most anitibiotic known to date are products of streptomyces and some fungi.

Advantages in using antibiotics

- Easy to produce antibiotics at cheaper rate.
- Cure incurable diseases.
- Less number of sprays required.
- Less leaching effect.
- Acts as eradicant.
- Manage MLO's and bacterial diseases.

Disadvantages in using antibiotics

- Shelf life very short.
- Less stable in storage.
- Narrow site of action.
- Risk of development of resistance.
- Only a few antibiotic are available for plant disease control.

Classification of antibiotics

I. Based on their activity

a) **Broad-spectrum:** Are active against a large number of microorganisms

E.g. Tetracylines, Chloromycetin etc.

b) **Narrow-spectrum:** Are effective against specific organisms only.

E.g. Penicillin.

II. Based on Chemical Group

1. β-Lacta antibiotics

Ex:

a) Penicillins

b) Cephlosporins

2. Tetracyclins

Ex:

a) Tetracyclins

b) Chlorotetracyclin

c) Oxytetracyclins

3. Rifamycins

Ex: Rifamycin A to E

4. Amino glycosidic antibiotics

Ex: Streptomycin, Gentamycin

5. Macrolide antibiotics

a) Erythromycin

b) Spyramycin

c) Olendomycin

6. Polypeptide group antibiotics

Bacitracine, Capriomycine, Polymixin, Viomycine

7. Miscellaneous group

Vancomycine, Chloromphenical, Fusidic acid, Lincomycin

8. Synthetic antibiotics

Ex: Nitrofuron compounds

a) Furazolidin - Stomach

b) Nitrofurantoin - Urinary track infection

c) Nitrofurozone - Ear infection

Important antibiotics and their mode of action

Name	Source	Effective against	Mode of action
Penicillins	*Penicillium* spp.	Prokaryotes	Inhibits murein synthesis
Cephalosporins	*Cephalosporium* spp.	Prokaryotes	Inhibits murein synthesis
Streptomycin	*Streptomyces griseus*	Prokaryotes	Binds to protein S12 of 30S causing aberrant inhibiton complex
Kanamycin	*S. kanamyceticus*	Prokaryotes	Affect 30S risbosomal aubunits and prevents the translation.

Antibiotic	Source	Target	Mode of action
Neomycin	*S. fradiae*	Prokaryotes	Affect 30S risbosomal aubunits and prevents the translation.
Erythromycin	*Streptomyces erythraeus*	Prokaryotes	Inhibits translation by binding to the 50S ribosome
Carbomycin	*S.halstedii*	Prokaryotes	Inhibits translation by binding to the 50S ribosome
Chlorotetrac-yclin	S. aureofaciens	Prokaryotes	Inhibits binding of aminoacyl-t-RNA to A site of 30S ribossomes; specificallybinds to the protein S& near A site, resulting change in topology
Oxytetracy-cline	S. riomosus	Prokaryotes	Inhibits binding of aminoacyl-t-RNA to A site of 30S ribossomes; specificallybinds to the protein S& near A site, resulting change in topology
Rifampin	*Amycolatopsis rifamycinica*	Prokaryotes	Inhibits DNA dependent RNA polymerase in bacterial cells by binding its betasubunit, thus preventing transcription to RNA and subsequent translation to proteins.
Bacitracin	*B. subtilis*	Prokaryotes	Destroys murein biosynthesis
Polymixin-G	*Bacilus polymyxa*	Prokaryotes	Destroys cytoplasmic membrane
Chloramphenicol	*Streptomyces venezuelae*	Prokaryotes	Inhibits peptidyl transferase activity of 50S ribosome & affects translation.
Nystacin	*S. nouresii*	Eukaryotes (Fungi)	Inactivates membrane containing sterols by making pore through which kions, small molecules.
Amphotericin	*S. nodosus*	Eukaryotes (Fungi)	Inactivates membrane containing sterols by making pore through which kions, small molecules.
Aureofungin	*Streptoverticillium cinnamoneus var terricola*	Eukaryotes (Fungi)	Inactivates membrane containing sterols by making pore through which kions , small molecules.
Natamycin/ Pimaricin	*Streptomyces natalensis*	Eukaryotes (Fungi)	Inactivates membrane containing sterols by making pore through which kions, small molecules.

31

Formulations of Fungicides and Antibiotics

Fungicide Formulation

Fungicide formulation is the process by which the fungicide is put into a form which can be easily produced, stored, transported and applied by practical methods in order to achieve a safe, convenient economic and effective method of disease control. The mixture of active ingredient and inert (inactive) ingredients is called a fungicide formulation.

Objective of Formulation

- To uniformly spread a small amount of the active ingredient over a large area.
- To ensure safety in handling and application.
- To optimize funicide doses for better efficacy.

Types of fungicide formulations

1. Solid formulation
2. Liquid formulation

1. Solid formulations

a) **Dusts (D)**: Free flowing powers contain technical material in the range of 2 to10% inert carrier.

b) **Granules (G)**: Genuine products contain technical material in the range of 3 to 10% and granule base.

c) **Water Dispersible Powders (WP)**: Free flowing powders containing technical material in the range of 25 to 75% wetting, dispersing agents and carrier.

d) **Soluble powders (SP)**: Soluble powders provide most of the same benefits as wettable powders, without they need for agitation once they are dissolved in the spray tank.

2. Liquid formulation

a) **Emulsifiable Concentrates (EC)**: Liquid formulations based on technical material which are not soluble in water contain 25 to 80% technical material solvent and emulsifier.

b) **Water Soluble Liquids (SL)** : Liquid formulation based on technical material which are soluble in water contain 36 to 85% technical material and solvent.

c) **Suspension Concentrates** : Thick slurry like formulations mixable in water.

d) **Aerosols** : They are liquids under pressure filled in cans which on release give a misty spray.

e) **ULV Formulations**: These are liquid formulations suitable for ultra low volume application.

Disadvantages of Conventional Formulations	Advantages of New Generation Formulations
Bulky, dusty and inconvenient	Improved residual activity
Hazardous during manufacturing, packing and application	Longer application intervals
Highly flammable due to use of organic solvents	Reduction in application dosage
Can cause phytotoxicity	Reduction in spray drift
More impact on non target organism	Less impact on non target organisms
Poor rain fastness	Better rain fastness
Corrosive to metal and plastic	Less phytotoxicity
Expensive to pack and transport	Constant and delayed biological effect
Dermal hazards	Reduced environmental pollution
Extreme inhalation danger	Reduced volatilization and leaching
	Safe storage due to reduced flammability

Fungicides and antifungal antibiotics

Fungicide Group	Name of the Fungicide	Formulation Registered
Inorganic sulphur	Sulphur	80% WP, 40% SC, 80% WG
	Lime sulphur	22% SC
Dithiocarbamates	Ferbam	75% W.P
	Mancozeb	75% WP, 35% SC
	Propineb	70% WP
	Thiram	80 WP
	Zineb	80% WP, 27% CS
	Ziram	80% WP, 27% CS
Copper based	Copper hydroxide	77% WP
	Copper orxychloride	50% WP, 40% Paste, 5% DP, 50% WG
	Copper sulphate	Tech 8%, 25% w/w, 2.62% SC
	Cuprous oxide	4% DP

Pthalimides	Captan	50% WP 75% WP
	Captafol	80% DS
Crotonates	Dinocap	48% EC
Organomercurial	Methoxy ethyl mercury chloride	3% FS, 6% FS
Phathalonitrites	Chlorothalonil	75% WP
Guanidine drivatives	Dodine	65% WP
Quinones	Diathianon	75% WP
Oxanthins	Carboxin	75% WP
	Oxycarboxin	20% EC
	Carboxin 37.5%+Thiram 37.5%	75% WP
Benzimidazoles	Benomyl	50% WP
	Carbendazim	25% DS, 50% WP, 46.27% SC
	Carbendazim 12%+ Mancozeb 63%	75% WP
Thiophanates	Thiophanate methyl	79% WP
Phosphorothiolates	Edifenphos	50% EC
	Iprobenfos	48% EC, 17% Gr
Dicarboximides	Iprodione	50% WP
Acylanines	Metalaxyl	35% WS,40% WS
	Metalaxyl 8%+ Mancozeb 64%	72% WP
Cyano-acetamide	Cymoxanil	80% WP
oximes	Cymoxanil 8%+ Mancozeb 64%	72% WP
Alkyl phosphonates	Fosetyl –Al	80% WP
Triazoles	Propiconazole	25% EC
	Peconazole	10% EC
	Myclobutanil	36% SL
	Triadimefon	25% WP
	Bitertanol	25% WP
	Hezaconazole	5% EC, 5% SC
	Difenoconazole	25% WP
	Tebuconazole	2% DS
Morpholines	Tridemorph	80% EC
Pyrimidines	Fenarimol	12% EC
Melanin biosynthesis	Tricyclazole	75% WP
inhibitor	Isoprothiolane	40% EC
Antifugnal antibiotics	Aureofungin	46.1% SP
	Kasugamycin	3% SL
	Validamycin	3% L

Formulations of Antibiotics

Name of the Antibiotic	Formulation registered
Actispray	3% cyclohexamide, soluble tablets
Agrimycin 100	7.7% cyclohexamide soluble tablets
Agrimycin 500	15% Streptomycin sulphate, 1.5% terramycin
Agristrep	1% Streptomycin sulphate, 1% terramycin + basic copper sulphate (1% copper)
Phytomycin	Streptomycin
Streptomycin	Streptomycin nitrate, 20% aqueous solution
Terramycin	Derived from *Streptomycin griseus*

32

Applications of Fungicides for Disease Management

Proper selection of a fungicide and its application at the correct dose and the proper time are highly essential for the management of plant diseases. The basic requirement of an application method is that it delivers the fungicide to the site where the active compound will prevent the fungus damaging the plant. The fungicidal application varies according to the nature of the host part diseased and nature of survival and spread of the pathogen. The method which are commonly adopted in the application of the fungicides are discussed

1. Soil application
2. Seed treatment
3. Foliar or vegetative application

1. Soil application

The purpose of soil treatment with chemicals is to reduce or eradicate the inoculums density of the pathogens present in *soil* as *soil invaders* or as *soil inhabitants.* The chemical soil treatment can be done in any of the following ways.

a) **Soil drenching**: Fungicidal suspension or solution is applied to soil surface in quantities enough to wet 10-15 cm depth of the soil before or after planting.

b) **Broadcasting**: Non-volatile fungicides (dusts, powders, granules, etc.) are mixed with soil or fertilizers and spread on the soil surface. Light ploughing or harrowing is then done to mix the chemical in sufficient depth.

c) **Furrow application**: In this method fungicides are applied either as dusts or with water to the furrow at the planting time. This method is possible in crops planted in furrows/rows such as potato and sugarcane. This method is good for control of diseases that occur at the base of the plants. This method requires much less quantity of the chemicals per hectare than the broadcast method.

d) **Fumigation** This method is usually used to control plant parasitic nematodes. The chemicals used are generally volatile and on coming in contact with soil moisture release gases which diffuse in the soil and kill the nematodes. Special equipments (Guninjectors) are required for application since diffusion is restricted to certain distance. The chemicals are injected into soil at regular distance (12-15 inch) all over the field. The depth of application is kept at 6-9 inches.

2. Seed treatment

Seeds, tubers, bulbs and other propagating materials are often treated with fungicides for eradication of the pathogen propagules present on and/or in them as well as for preventing their rot in the soil after planting. The chemicals can be applied on the seed as dusts, or as thick water suspension (slurry) mixed with the seed. The seeds can also be soaked in water or solvent solution of the chemical and then allowed to dry. On the basis of their mode of action seed treatment chemicals can be of three types:

a) **Seed disinfestants**: Those which kill the pathogen present on the seed and do not remain active for long after the seed has been planted.

b) **Seed disinfectants**: Those which act as eradicants and destroy the pathogen established in seed tissues. A number of systemic chemicals are being used for this purpose.

c) **Seed protectants**: When seed are treated with protective fungicides, the pathogens present on the seeds are destroyed and the compounds remain on seed surface for sufficiently long times (In treating seed and other propagative organs, it should be ensured that enough chemical has sticked to the seeds. It should also be ensured that their viability is not lowered or destroyed.

d) **Seed biopriming:** *Treating of seeds with biocontrol agents and then incubating under warm and moist conditions until just prior to emergence of radical is reffered as bioprimming.* This technique has potential advantages over simple coating of seeds as it results in rapid and uniform seedling emergence. *Trichoderma* conidia germinate on the seed surface and form a layer around bioprimed seeds. Such seeds tolerate adverse various soil conditions better. Biopriming could also reduce the amount of biocontrol agents that is applied to the seed.

3. Spraying

This is the most commonly adopted method. Spraying of fungicides is done on leaves, stems, flowers and fruits. Wettable powders are most commonly used for preparing sprays. The most common diluent or carrier is water. The dispersion of the spray is usually achieved by its passage under pressure through nozzles of sprayers. Spraying is of two types, namely, *high volume* and *low volume.* When sprays involve large quantities of liquid per unit area, they are termed *high volume. Six hundred litres and above per hectare* would be considered to be of the *high volume* category. With *low volume* sprays, it is usually possible to cover *one hectare* with about *100 litres or less.*

4. Dusting

Dusts are applied to leaves, fruits and stems of plants as an alternative to spraying. Dry powders are used for covering host surface. *Dusting* is practicable only in the *calmest weather* and high effectiveness is obtained if the dust is applied when the plant is *wet with dew or rain.*

5. Paste or Paints

In orchards, the wounds created during pruning and training of trees often serve as entry points of the pathogens. To protect these wounds fungicidal *pastes* or *paints* are used as protective layer. These pastes are prepared with fungicides in a suitable carrier such as raw linseed oil, lanolin, glycerine, etc. the residual effect of the pastes lasts long enough to permit natural healing of the cut surfaces

Questions

Short answer

1. What is the mode of action of a protectant fungicide?
2. Why protactant fungicides usually provide long term and durable control in comparison to systemic fungicides?
3. What are the main factors behind pathogen resistance to fungicides?
4. Mention difference between protectant and systemic fungicides?
5. Why 1960s is considered to be the turning point in the area of chemical treatment of plant diseases?
6. Who discovered Bordeaux mixture? What features make bordeaux mixture an excellent fungicide?
7. What are the fundamental differences between a systemic fungicide and an antibiotic?
8. Name some important antibiotics used for plant disease management?
9. Who discovered the first dithiobarbamate fungicide? What name was attributed to this fungicide.

10. Mention important features of an ideal systemic fungicide.
11. Name two acylaline fungicides used effectively against omycetous fungi?
12. What is antibiotic? Write down advantage and disadvantage of antibiotic in plant disease management.
13. Antibiotics in plant disease management.
14. Application technology.
15. Milestones in the development of fungicides

Long answer

1. Write short notes on the following ;

a) Sterol biosynthesis inhibitors

b) Strobilurins

c) Acylalines

d) Oxantiins

e) Organo tin compounds

2. What are systemic fungicides ? Give a detail account of systemic fungicides in plant disease management .
3. What is soil fumigations ? Discuss the role of important soil fumigants in plant disease management.
4. Discuss various application patterns of chemicals in plant disease managment.
5. Gives various methods of classification of fungicides. Support your statement with suitable examples.
6. Enlist important fungicides belonging to organic sulphur group. How mancozeb is toxic to fungi.
7. Write short notes on: a. Characteristic of an ideal fungicide b. Factors affecting the performance of a fungicide.
8. How Bordeaux mixture is prepared ? Describe method, chemical reactions, precaution and advantage of Bordeaux mixture.

33

Integrated Disease Management

Definitions

Selection, integration and implementation of disease control means on predicted economic, ecological, and sociological consequences.

A comprehensive approach to disease control that uses combined means to reduce the status of diseases to tolerable levels while maintaining a quality environment.

The optimization of disease control in an economically and ecologically sound manner, accomplished by the coordinated use of multiple tactics to assure stable crop production and to maintain disease loss below the economic injury level while minimizing risks to man and the environment.

A sustainable approach to manage diseases by combining biological, cultural, physical and chemical tools in a way that minimizes economic, health and environmental risks.

Goals of IDM

The main goal s of an integrated plant disease management programme are to

- Eliminate or reduce the initial inoculums
- Reduce the effectiveness of initial inoculums
- Increase the resistance of the host
- Delay the onset of the disease
- Slow the secondary cycles

Objective of IDM

Varied IDM objective are:

- To reduce potential hazards to environment and people health.
- To provide improved control by conservation of natural enemies and employing traditional method.
- To help in the production of improved quality produce.

- To reduce the cost of plant protection.
- To help in improvement on water and soil quality.
- To reduce farmers and consumer risk.
- To facilitates better pesticide management.

Why practice IDM

There are many reasons to broaden disease management beyond the use of chemicals

1. **IDM helps to keep a balanced ecosystem**: Every ecosystem has a balance, the actions of one kind of organisms in the ecosystem usually affect other species. Introducing chemicals into the system can change the balance, destroying certain species and allowing other species to dominate. Pesticides can kill beneficial microorganisms that consume pathogens, leaving few natural mechanisms of disease control.
2. **Pesticides can not always effective**: Chemical pesticides are not always effective. Pathogen can become resistant to pesticides. Furthermore, pathogens may survive in situations where the chemical does not reach pathogens, is washed off, is applied at improper rate, or is applied at an improper life stage of pathogen.
3. **IDM can save money**: IDM can avoid crop loss and prevent unnecessary pesticides expense. Applicators can save on pesticide costs because the need for control rather than the calendar is the basis for applying pesticides.
4. **IDM promotes a healthy environment**: We have learned about the persistence of chemicals in the environment and their effects on living creatures. Cases of contaminated ground water appear each year and disposal of containers and unused pesticides still pose challenges for applicators. Make sure that environmental impacts are considered in any disease management decisions. Using IDM strategies helps keep unreasonable adverse effects to a minimum.
5. **IDM maintains a good public image:** IDM is now demanded by many sectors of our society. IDM has been implemented to grow our food, to protect crops from diseases and pests and to protect humans and livestock health.

Component of IDM

An IDM programme is designed around six basic components.

1. **Acceptable disease levels:** The emphasis is on management, not eradication. IDM holds that wiping out an entire disease is often impossible

and the attempt can be more costly, environmentally unsafe, and frequently unachievable. IDM programs first work to establish acceptable disease levels, called action thresholds, and apply controls if those thresholds are crossed.

2. **Monitoring:** Regular observation is the cornerstone of IDM. Observation is broken into two steps. first; inspection and second; identification. Visual inspection, other measurement methods and monitoring tools are used to monitor disease levels. Accurate disease identification is critical to a successful IDM program. Record-keeping is essential, as is a thorough knowledge of the behavior and disease cycles of target diseases. Monitor the degree days of an environment to determine when is the optimal time for a specific disease' outbreak.
3. **Genetic host resistance**: The use of genetically resistant plants is often recommended by plant pathologists as the first line of defense to minimize yield losses due to diseases. In some cropping systems, such as large acreage field crop like corn, soybeans wheat, rice and cotton, the use of genetically resistant plants may be the only cost-effective means for managing a particular disease. In some cases, the use of resistant cultivars or varieties might be the only means of effectively managing a disease such as in the case of managing plant diseases caused by viruses. The development of resistant plant types may also reduce the need for using pesticides. Although genetic resistance should be considered when dealing annual cropping systems where new seed is sown each season thereby providing an opportunity to introduce new cultivars or varieties with pathogen resistance. Although important in perennial cropping systems such as orchards, forests, golf courses, or home lawns, once the initial crop is planted, the introduction of resistant lines is limited due to the long-term nature of these crops.
4. **Cultural practices:** Cultural practices not only serve in promoting the healthy growth of the crop, but are also effective in directly reducing inoculums potential and enhancing the biological activities of antagonists in the soil. Cultivation practices such as cultivation techniques, intercropping, planting date, plant density, crop rotation, roughing, healthy planting material, soil solarization, soil amendments, water management, fertilizer management have been used singly and in combination as tools for disease management. For some crops, such control practices may be the only economically viable method available.
5. **Mechanical controls:** Should a disease reach an unacceptable level, mechanical methods are the first options to consider. They include simple hand picking, hot water and steam treatment, soil solarization and tillage to disrupt breeding.

6. **Biological controls:** Natural biological processes and material can provide control, with minimal environment impact, and often at low cost. The main focus here is on promoting bioagents that control target diseases. Biological fungicides derived from naturally occurring microorganism (e.g Trichoderma and Pseudomonas) also fit into this category.
7. **Chemical controls:** Fungicides played an important role in disease control for many decades. The contact fungicides such as mancozeb, chlorothalonil, copper and sulphur based products continued to have a good share of the market in India because of their lower cost.. However, systemic fungicides started gradually to replace the older non systemic chemicals with more effectiveness and specificity in disease control. Very quickly, triazoles fungicides gained popularity. Of late, new classes of fungicides were developed with significant impact on disease control. These include anilinoprimidines, phenoxyquinolines, oxazolidinediones, spiroketalamines, phenylpyrroles, strobilurins and activators of systemic acquired resistance. The availability of a variety of new products , with narrow and broad specificity , offer important disease control options, however, their practical application continues to face the risk of selection of resistant pathogen populations. The fungicidal application had a better impact when used within an IDM strategy. The fungicide industry is focusing in developing effective products with low mammalian toxicity and environmental impact and low residues in food, to meet international health standards and compatibility in IDM programs.

Use of disease level thresholds

Economic Injury Level: It is the lowest disease index that will cause economic damage

Economic Threshold Level: It is the disease index at which control measures should be applied to prevent an increasing disease from reaching the economic injury level

Action Threshold Level: An action threshold is the disease level at which some type of disease management action must be taken. It is a predetermined disease level that is deemed to be unacceptable.

Advantages of IDM

- Promotes sound structures and healthy plants.
- Promotes the sustainable bio based disease management.
- Reduces the environmental risk associated with management.
- Encourage adoption of more ecologically benign control tactics.

- Reduces the potential for air and ground water contamination.
- Protects the non-target species.
- Reduces the need for pesticides by using several management methods.
- Reduces or eliminates issues related to pesticide residue.
- Decreases workers, tenants and public exposure to chemicals.
- Maintains or increases the cost-effectiveness of disease management programs.

Constraints of IDM

- Unavailability of multiple resistant varieties.
- There is a need for a number of selective pesticides against diseases of crops which are safe against the natural enemies.
- Economic threshold levels have not worked out for most of the diseases.
- Precise disease surveillance and monitoring methods and forecasting models have not been standardized.
- Techniques of mass multiplication of several potential bio control agents are still not well developed.
- Lack of multi disciplinary approaches inter institutional collaboration to develop sound IDM technology.
- Lack of trained Human resource development in IDM.

Questions

Short answer

1. Mention two important problems encountered due to continuous and injudicious use of fungicides?
2. What is IDM?
3. Why IDM is considered to be a system approach?
4. Define EIL, ETL & ATL.
5. Why practice IDM?

Long Answer

1. What important phases significantly contribute in development of a complete IDM system?
2. What is IDM? Describe objective, advantages and disadvantages of IDM.
3. What is IDM? Describe component of IDM with suitable examples?

34

Multiple Choice Questions

1. The nematodes are –

   a) Monoblastic b) Duoblastic

   c) **Triploblastic** **d)** All of these

2. Bacterial cell division mainly by

   a) Binary fission b) Fragment

   c) Budding d) None of these

3. Rice blast pathogen perfect stage is–

   a) *Pyricularia oryzae* **b) *Magnaporthe grisea***

   c) Helmiuthosporium d) Rhizoctonia

4. Downy mildew of pea caused by–

   a) *Perenospora pisi* b) *Albugo candida*

   c) *Erysiphae polygoni* d) None of these

5. Black heart is a physiological disorder of

   a) Tomato b) Chilli

   c) Potato **d)** Cabbage

6. The scientific name of burrowing nematode is–

   a) *Xiphinema* sp. b) *Lougidorus* sp.

   c) *Meloidogune* sp. **b) *Radopholus* sp.**

7. The pathogen of rice blast or 'rotten neck' is–

   a) *Ustilago tritici* **b) *Pyricularia oryzae***

   c) *Alternaria altemata* d) None of these

8. Tea rust is caused by–

   a) MLO b) Virus

   c) Bacteria d) **Algae**

Bold means correct answer

9. The major storage fungi that effects the food grain is–

a) Rhizobium b) Mucor

c) Cercospora d) **Aspergillus**

10. Which one of the following microbial agents is being commercially exploited as biocontrol agent ?

a) *Xanthomonas campestris* b) *Penicillium notatum*

c) *Sclerotium rolfsii* d) ***Trichoderma viride***

11. Which Penicillium species used in 'Cheese' making ?

a) *Penicillium notatum* b) ***Penicillium roqueforti***

c) *Penicillium diuaricatum* d) *Penicillium crysogemum*

12. Application of potash increases

a) Resistance to water logging b) Frost resistance in plants

c) Disease resistance in plants d) None of these

13. Viruses contain–

a) RNA b) DNA

c) Both RNA and DNA **d) Either RNA or DNA**

14. The incidence of black scurf of potato is more in–

a) Sandy soil b) Clay soil

c) Alluvial soil d) Loam soil

15. Teliospore of rust have germpores in number–

a) 1 b) **2**

c) 3 d) 4

16. In 2005 which pathological scientist got Borlaug award–

a) Rattan Lal b) VL Chopra

c) CD Mayee d) **S Nagarajan**

17. The downey mildew disease of pearl millet is primarily a–

a) Seed borne b) Air borne

c) Soil borne **d)** Water borne

18. Renowned scientist Dr. N.E. Borlaug belongs to–

a) Agronomy b) Genetics

c) Entomology d) **Plant Pathology**

Bold means correct answer

19. Select the nematicides group which are non-fumigant in nature–
    a) Methyl bromide b) DBCP
    c) Vapam d) **Aldicarb and Phorate**
20. The causal organism of bunchy top of banana is transmitted by–
    a) *Pentalonia nigronervosa* b) *Bemisia tabaci*
    c) *Lipaphis erisimi* d) Pollen
21. Sulphur dioxide is toxic to plant at or above–
    a) 0.1-0.3ppm b) **0.3-0.5ppm**
    c) 0.5-0.7ppm d) 1 ppm
14. Damping off and leaf blights are very effectively checked by–
    a) Bordeaux mixture b) Burgundy mixture
    c) Thirarn d) **Copper oxychloride**
15. Yon Schmeling and Kulka first time demonstrated the systemic activity of fungicides in–
    a) 1956 b) **1966**
    c) 1972 d) 1976
11. The powdery mildew of wheat is associated with fungi–
    a) *Enysiphe graminis sp. tritici* b) *E. graminis sp. hordei*
    c) *Enjsiphe polgoni* d) None of these
12. Most widely used fungicide for rust fungi is–
    a) Vitavax b) **Plantvax**
    c) Bavistin d) Dithane M -45
13. Who is the father of mycology ?
    a) EJ Butler b) **PA Michelli**
    c) Woronine d) Anton de Bary
14. The science which deals with cause of the plant diseases and nature of the causal agent–
    a) Fertilization b) Etiology
    c) Aetiology d) **Both B) & (C)**
15. Select the disease in which small droplets of pinks or light honey coloured–fluid (the honeydew stage) exudes from the spikelets–
    a) Citrus canker b) Green ear diseases of bajra
    c) Ergot of bajra d) Bacterial blight of rice

Bold means correct answer

16. Dinocap is sold in market as–
    a) Bravo b) Dexon
    c) Botron d) **Karathane**
17. Mad cow disease is caused by
    a) Virion b) **Pirion**
    c) Bacteria d) MLO
18. Race specific resistance called–
    a) Vertical resistance b) Horizontal resistance
    c) Adult plant resistance d) Apparent resistance
19. 'Buck eye rot' is a disease of which crop
    a) Water chestnut **b)** Sweet potato
    c) Pods of garden pea d) Tomato fruits
20. Sterility mosaic disease of pigeonpea spread by
    a) Virus b) Aphid
    c) Whitefly d) **Mites**
21. Potato spindle tuber disease is transmitted–
    a) Mechanically b) Biologically
    c) Water d) All of these
22. Potato virus diseases are spread by –
    a) Aphids **b)** Jassids
    c) Nematodes d) Tuber moth
23. The cauliflower mosaic virus particles contain–
    a) RNA **b) DNA**
    c) Both (a) and (b) d) Either RNA or DNA
24. In north India and central India, the black rust inoculums cause from–
    a) South b) Hilly area
    c) From USA d) From Nepal
25. Decomposition of organic matter in submerged soil is carried out by–
    a) Bacteria b) Actinomycetes
    c) Fungi d) Earthworm

Bold means correct answer

26. Yellow leaf mosaic of bhindi crop spreads by–

a) Jassids
b) Borers
c) Jassid and borers
d) None of these

27. The Bordeaux mixture was discovered by Millardet at Boardeaux in France in year–

a) 1807
b) **1885**
c) 1905
d) 1936

28. Who did most of his work on rust diseases in India–

a) R Prasad
b) **KC Mehta**
c) BB Mundakur
d) EJ Butler

29. The plants produce a whip like black shoot, often very long and much curved–on itself in smut of sugarcane is caused by pathogen–

a) *Ustilngo nuda*
b) *Ustilngo tritici*
c) *Ustilago sitanminea*
d) *Ustilngo hordei*

30. The fungi which transmit plant viruses belong to class–

a) Basidiomycetes
b) Oomycetes
c) Zygomyctes
d) **Plasmodiophoromycetes**

31. The green ear or downy mildew of pearlmillet was first time reported in India by–

a) KC Mehta
b) **E J Butler**
c) BB Mundakur
d) R. Prasad

32. Black rust or stem rust of wheat is caused by –

a) *Puccinia graminis tritici*
b) *Puccinia strriformis*
c) *Puccinia recondita*
d) *Melampsora lini*

33. Which systemic fungicide was first time discovered by Von Schmeling and Kulka?

a) Oxathiin
b) Pyrimidines
c) Benzimidazoles
d) Quinones

34. All plant viruses (except cauliflower mosaic virus) contain–

a) DNA
b) **RNA**
c) Both RNA and DNA
d) Either RNA or DNA

Bold means correct answer

35. Sucidal germination takes place in–

a) Dodder b) **Striga**

c) Lorattthus d) *Dendrophthoe falcata*

36. Cell wall lack microorganism

a) Bacteria b) Virus

c) Fungi d) **MLO**

40. Citrus greening caused by the–

a) Fastidious bacteria b) Fungi

c) Virus d) MLO

41. The pathogen responsible for charcoal rot of soybean is–

a) *Ascochyta rabiei* **b) *Macrophomina phaseolina***

c) *Rhizoctonia solani* d) Penicillium

42. The effective control of flag smut of wheat (*Urocystis tritici*) can be done by adopting of–

a) Use of resistant varieties b) Seed treatment

c) Crop rotation d) **All of these**

43. Karnal bunt of wheat first time reported in Karnal by Mitra in–

a) 1929 b) 1931

c) 1941 d) **1951**

43. Smut of maize caused by–

a) *Ustilago tritici* b) ***Ustilago maydis***

c) *Ustilago hordei* d) None of these

44. The *Colletotrichum falcatum* produces –

a) Zygospores b) Oospores

c) Ascospores d) **Conidiospores**

45. Infection of smut of maize (*Ustilago maydis*) occurs during–

a) Vegetative stage b) Reproductive stage

c) After flowering d) Before sowing

46. Citrus canker (lesions), which originated from China is caused by pathogen

a) *Xanthomonas campestris pv. citri*

b) *Albugo candida*

c) *Enoinia annjlouora*

d) *Claviceps fusifomis*

Bold means correct answer

47. The ascus is a sexual fruit of ascomycetes having a globose shape but no ostiole (opening) is known as–

 a) Cleistothecia **b)** Perithecia

 c) Apothecia d) All of these

48. 'Little leaf' in brinjal is caused by a–

 a) Fungus b) Bacteria

 c) Virus **d) Mycoplasma**

49. The viruses which are usually helped or accompanied by smaller spherical particles of another serologically unrelated virus known as–

 a) Satellite virus b) Geminiviruses

 c) Virid d) Capsid

50. 'White blisters of crucifers' is caused by pathogen–

 a) *Pythium debarnum* b) ***Albugo candida***

 c) *Sclerospora sorghi* d) *Plasmopara viticola*

51. Early blight of potato produces –

 a) Conidia b) Telia

 c) Uredia d) Acecia

52. Who is the father of American Nematology ?

 a) Kuhn b) H Schacht

 c) NA Cobb **d)** T J Burill

53. Brown leaf spot disease of rice is caused by–

 a) *Drechslera oryzae* b) *Cercospora arachidicola*

 c) *Xanthomonas onyzae* d) *Pyricularia oryzae*

54. The nematode *Heterodera avenae* is confined to the gramineae family except one single legume–

 a) *Pisum sativum* b) *Cajanus cajan*

 c) *Vigna mungo* d) ***Senebiera pinnatifolia***

55. The effective control of late blight of potato is possible by use of–

 a) Sanitation measures b) **Spray of metalaxyl**

 c) Bordeaux mixture d) All of these

56. Most widely used biocontrol agent is

 a) *Pseudomonas florescence* **b)** *P. putida*

 c) *Bacillus subiilis* d) Clostridium

Bold means correct answer

57. Yellow mosaic of legumes was first reported in India from–
   a) Shimla b) Solan
   c) Delhi d) Kanpur
58. MLD and spiroplasma are mostly–
   a) Xylem inhibiting **b) Phloem inhibiting**
   c) Both a)and b) d) Stomata inhibiting
59. Which fungicide is used against downey mildew?
   a) Metalaxyl **b)** Karathane
   c) Oxathin d) Streptomycin
60. MLO disease transmitted by–
   a) Leaf hopper b) Aphid
   c) Whitefly d) Animals
61. Nematicides inhibits which enzyme of nervous system in killing of nematodes–
   a) Kinase b) Isomerase
   c) Phosphatase d) **Acetyl cholinesterase**
65. Virus capsid is made up of–
   a) Protein **b)** Carbohydrate
   c) Lipid d) Nucleic acid
66. The term virus was first time coined by
   a) Kuhn b) **Beijerinck**
   c) Stanley d) None of these
67. In rust cycle the cereal host is infected by–
   a) Urediospores b) **Aeciospores**
   c) Teliospores d) Basidiospores
68. The attack of sheath blight of rice caused by *Rhizoctonia solani* is more during–
   a) Germination **b) Active tillering stage**
   c) Flowering d) All of these
69. Karnal bunt of wheat gives foul smell in the field due to presence of volatile compound–
   a) Tetramethyl amine b) **Trimethyl amine**
   c) Diallyl propyl sulphide d) Allyl propyl disulphide

Bold means correct answer

70. Father of Indian Mycology–

a) EJ Butler b) KC Mehta

c) BB Mundakur d) RS Singh

71. Common scab of potato can be controlled by reducing pH below 5.2 is caused by the pathogen–

a) *Streptomyces scabies* b) *Streptomyces griseus*

c) *Streptomyces aureofaciens* d) *Streptomyces flaueolus*

72. In 1845, the late blight of potato destroyed the potato crop of Ireland was caused by–

a) *Phytophthora infestans*

b) *Alternaria solani*

c) *Pythium aphanidermatum*

d) *Pseudomonas solanacearum*

73. The nematodes lack organs for–

a) Circulation b) Respiration

c) Both a) & b) d) Excretory

74. In the attack of which class of nematodes, the presence of white to brown cysts projecting on the root surface is a characteristic symptom–

a) Longidorus b) **Globodera**

c) Meloidogyne d) Trichodorus

75. 'Plant Pathology' written by–

a) RS Singh b) **GN Agrios**

c) VS Singh d) AP Sinha

76. Hot water treatment of seed is useful for control of–

a) Loose smut **b)** Covered smut

c) Rust d) Powdery mildew

77. Rust includes in–

a) Deuteromycotina b) **Basidiomycotina**

c) Ascomycotina d) Oomycetes

78. The fungus *Drechslera oryzae* produces toxins which are highly toxic to rice seedling, name of such toxin is–

a) ABA b) Trimethyl amine

c) Isobutylene d) **Cochliobolin**

Bold means correct answer

79. Bacterial diseases are controlled by use of chemicals–

a) Kelthane b) Fungicide

c) Antibiotics **d)** Viricides

80. The symptoms of powdery mildew of pea caused by *Erysiphe polygoni* first appears on–

a) Stem b) Roots

c) Leaves d) Flowers

81. MLO first discovered by–

a) Louis Pasteur b) Kuch

c) Doi *et al.* **d)** Anupam Verma

82. The cell wall of the mycelium of true fungi (Eumycota) contains

a) Glucan + cellulose b) **Glucan + chitin**

c) Glucan d) Peptidoglycon

83. Parasexuality was first discovered by

a) Pontecarvo b) Erikson

c) De Bary d) Robert Koch

84. An example of virus containing ds DNA

a) Caulimovirus b) Gemini virus

c) Tobravirus d) Potyvirus

85. Multiplication of virus occurs in

a) Epidermal cells b) **Plasmodesmata**

c) Phloem vessels d) xylem vessels

86. Sugarcane ratoon stunting is caused by

a) Phytoplasmas b) Viroids

c) ***Cavibacter xyli sub sp xyli*** d) Virus

87. Hyperplasia refers to

a) **Increased cell division** b) Cell death

c) Cell enlargement d) Decreased cell division

88. An internally seed borne fungal disease

a) Smut of Ragi b) Red rot of Sugarcane

c) **Loose smut of Wheat** d) Grain smut of Jowar

Bold means correct answer

89. Significant contribution on the epidemiology of disease cycle of stem rust of wheat was made by

a) **Mehta. K.C** b) Mundkur. B.B

c) Payak.M.M d) Nagarajan

90. Phytoplasmas contain

a) RNA only b) DNA only

c) **RNA+ DNA** d) RNA or DNA

91. MLO are mostly

a) Xylem inhibiting b) **Phloem inhibiting**

c) Both (a) and (b) d) Stomata inhibiting

92. NEPO viruses are transmitted by

a) *Xiphinema index* **b) Trichodorus sp**

c) *Pythium ultimum* d) Meloidogyne

93. Potato spindle tuber disease is caused by

a) Viroid b) Phytoplasma

c) Spiroplasma d) Virus

94. The Indian Phytopathological society was started by

a) B.B.Mundkur b) Coleman

c) E.J.Butler d) Kirtikar

95. Host plant on which wheat rust fungus produces telial stage is known as

a) Alternate host b) **Primary host**

c) Collateral host d) Secondary host

96. NETU viruses are transmitted by

a) Xiphinema **b)** Trichodorus

c) Longidorous d) Paralongidorus

97. Stylet-borne viruses are

a) Non persistent b) Circulative

c) Persistent d) Propagative

98. Who advanced the gene for gene concept of disease and resistance and susceptibility

a) Biffen and Orton b) **Flor**

c) Vander plank d) Gaumann

Bold means correct answer

99. Most of the Plant viruses contain

a) RNA b) DNA

c) Both RNA + DNA d) None

100. The famous Irish famine of 1842 was caused due to the outbreak of

a) *Leaf spot of rice* b) Stem rust of Wheat

c) Late blight of potato d) Downy mildew of Grapes

101. The term virus was coined by

a) Adolf Meyer b) Stanley

c) **Beijerinck** **d)** Ivanowski

102. Bacteriophage is

a) Naked single stranded RNA particle

b) A Virus infecting bacteria

c) A virus infecting eukaryotic cell

d) Animal virus

103. An example of simple interest disease is

a) Late blight of potato **b) Wiit of pigeonpea**

c) Stem rust of wheat d) None of the above.

104. Whiptail disease of cauliflower is caused due to deficiency of

a) Boron b) Iron

c) Magnecium **d) Molybdnum**

105. Bacteriophage word was given by

a) Mayer **b) D'Herelle**

c) Hashimoto d) Doi *etal*

106 Progressive death of twigs and branches from their tip toward the trunk is known as—

a) Decay b) Blight

c) Wilt **d) Die back**

107. Cuscuta is a

a) Complete stem parasite **b) Complete root parasite**

c) Semi stem parasite d) Semi stem parasite

Bold means correct answer

108. Chocking of vascular bundles by fungal hyphae is observed in

a) Root rots **b) Wilts**

c) Anthracnose d) Downy mildew

109. Which pathogen is associated with the discovery of Bordeaux mixture

a) ***Plasmopara viticola***. b) *Phytophthora infestans*

c) *Helminthosporium oryzae*. d) *Uncinula necator*

110. Which of the following is a complete root parasite

a) Orbanche b) Striga

c) **Cuscuta** d) None of the above

111. Domestic quarantine exists on

a) Rice blast b) Red rot of sugarcane

c) **Wart disease of potato** d) Wilt of pigeonpea

112. The safest, economical and most effective method of plant disease management is

a) Chemical management b) Biological management

c) **Use of resistant varieties** d) Cultural management

Bold means correct answer